Raging Planet

Earthquakes, volcanoes, and the tectonic threat to life on Earth

Guire

BARRON'S

A QUARTO BOOK

Copyright © 2002 Quarto Inc.

First edition for the United States, its territories and
dependencies and Canada, published in 2002 by
Barron's Educational Series, Inc.

All inquiries to be addressed to:
Barron's Educational Series, Inc.
250 Wireless Boulevard
Hauppauge, NY 11788
http://www.barronseduc.com

Library of Congress Catalog Card Number 2001087636

International Standard Book Number ISBN
0-7641-1969-9

QUAR.RAPL

Conceived, designed and produced by
Quarto Publishing plc
The Old Brewery
6 Blundell Street
London N7 9BH

Senior editor Nicolette Linton
Art editor/designer James Lawrence
Assistant art director Penny Cobb
Text editors Gillian Kemp, Claire Waite Brown
Illustrators Julian Baker, Patrick Mulrey,
              Tony Walter-Bellue
Picture research Sandra Assersohn
Indexer Dorothy Frame

Art director Moira Clinch
Publisher Piers Spence

Manufactured by Regent Publishing Services Ltd,
Hong Kong
Printed by Leefung-Asco Printers Ltd, China

9 8 7 6 5 4 3 2 1

AUTHOR'S ACKNOWLEDGMENTS
For my parents, Audrey and the late John McGuire.

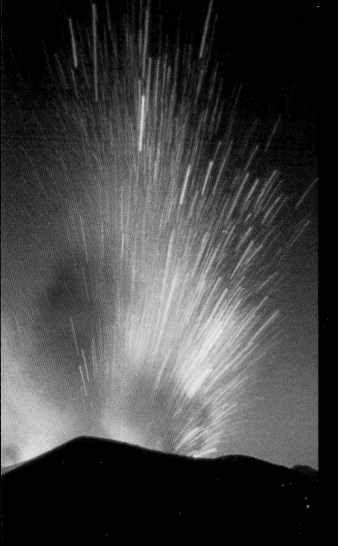

INTRODUCTION

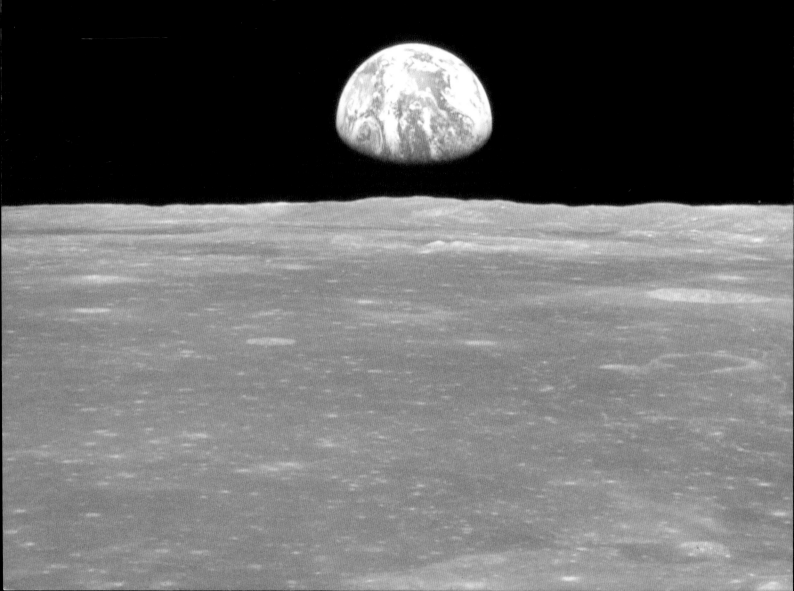

SMALL BLUE PLANET

If the Earth appears in a planet catalog of some advanced alien civilization, chances are, the gist of the accompanying description will read "innocuous and remote."

A small, rocky body orbiting a mundane star in a quiet backwater of the Milky Way galaxy, the Earth is nevertheless very special to us. It is, after all, the only home we have. Although our small blue planet appears serene and graceful as viewed by astronauts on the Moon or in the orbiting space station *Alpha*, this first impression disguises a dynamic planet that is seething with hidden energies. The Earth is a perilous cauldron under constant threat from both within and without—only a thin skin of cooled crust protects us from the boiling, molten maelstrom beneath our feet. The skin itself is far from stable: broken into chunks of rock that scrape and jostle against one another as they float on a semimolten layer. Where chinks appear, the molten rock from below bursts through to the surface, while in places huge slabs of rock founder and sink down into the baking interior. Periodically the rocky sphere is pounded by lumps of rock and ice from space as the Earth hurtles through a hostile solar system littered with debris.

As 6 billion souls can testify, sharing the most dynamic body in our family of planets with a constantly restless Mother Nature makes life far from easy. Although she is often benign, Mother Nature can explode in an orgy of violence against which we have little or no protection. So far she has prospered for over 4 billion years. Comparatively speaking, we have only been on the scene for the blink of an eye. We have,

perhaps, another 5 billion years before the Earth is swallowed up by a dying, bloated Sun —a terribly long time to coexist peacefully with a capricious Mother Nature. We must accept that the planet we live on and the space it flashes through may simply be too dangerous for us. Until now we have prospered, but the greatest battles with nature on this raging planet remain to be fought and the ultimate outcome may not be in our favor.

OPPOSITE The spectacular view of the Earth over the Moon, taken by the Apollo 8 astronauts in 1968. BELOW Rescuers fight to release survivors trapped by the 1995 earthquake on the Russian island of Sakhalin.

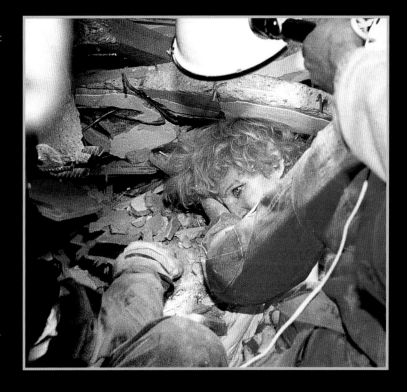

Earth: the early years

>>> The Sun and its entourage of planets, including our own Earth, are really the new kids on the block.

> While recent estimates suggest that it is now around 14 billion years since the biggest bang of all heralded the violent birth of our universe, our solar system began to come together a mere 4.6 billion years ago. The Sun, therefore, is far from being one of the founding fathers of space and time, and belongs to a younger generation of nature's fusion reactors fed by the atoms of much older stars that blew themselves apart in cataclysmic explosions billions of years earlier. This later-generation status is critical for the Earth and its sibling planets, because the most important elements in their make-up, especially iron, silicon, aluminum, and oxygen, were rustled up by nuclear wizardry in the later stages of the life cycles of older generation stars before being blasted into space during their final death throes.

In the beginning

Our solar system can trace its beginnings to a dense cloud of interstellar gas and dust that began to condense and shrink—due, some think, to the shuddering impact of shock waves from a nearby exploding star or supernova. As gravity took a hand, much of the debris was pulled together into a central mass which, within barely a million years, ignited to form a weakly shining proto-sun. Although the Sun's formation scavenged over 90 percent of the original cloud of gas and dust, plenty of material remained to form a rotating disc around the new star, rather like the rings we see around the planet Saturn. This flattened pinwheel of leftovers was to become the spawning ground of the solar system we know today. As the disc rotated, so individual particles within constantly collided with one another; at times sticking and coalescing to form larger particles and, at other times, disintegrating due to the energy of the collisions. On the whole, the tendency was toward the formation of fewer but larger bodies; a process known as accretion which, after some 100 million years, saw virtually all the Sun's leftover debris brought together into the nine planets we see now. The remainder went to form the planets' moons together with the rogue asteroids and comets that continue to threaten our planet today.

After its formation, the Earth was not, however, to find immediate tranquility. Bombardment by remaining bits and pieces of space debris continued, including one chunk the size of Mars. The force of that collision was almost sufficient to split the Earth apart. Fortuitously this did not happen, but a huge mass of debris was expelled that quickly regrouped to form our nearest celestial neighbor, the Moon. The collision may also have knocked the Earth off its axis, the feature of our planet that gives us the seasons. Like the roaming asteroids and comets, the Moon also plays a role in the hazardous present and future of our planet. Its gravity not only controls the ocean tides, but also tugs at the crust with sufficient strength to trigger earthquakes and volcanic eruptions.

By the end of the accretion process, the Earth was generating enormous amounts of heat due to the high concentration of radioactive elements in the interior, augmented by the

Crust

Mantle

Light silicates rise to form primitive crust

Dense metals, such as iron and nickel, sink to form core

Core

ABOVE Slowly, the different components of the Earth separated to form a metallic core enclosed in a rocky mantle and crust.

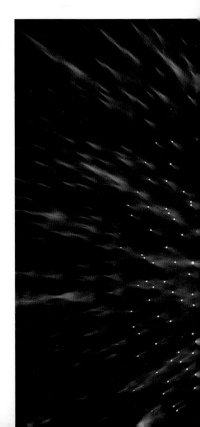

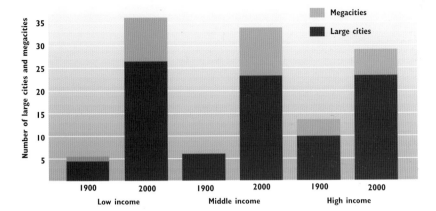

Number of large cities and megacities

Megacities
Large cities

| 1900 | 2000 | 1900 | 2000 | 1900 | 2000 |
| Low income | | Middle income | | High income | |

and over 140 supercities holding over 2 million people, most located in poor, developing countries in areas prone to natural disasters. In a few years, for the first time in history, more people will live in urban environments than in the countryside, and, by 2025, over 5.5 billion people will be crammed into gigantic cities—more than the population of the planet in 1990.

Many of these huge urban agglomerations—such as Mexico City and Tokyo—are located close to plate boundaries or other tectonically active areas, where destructive earthquakes and volcanic eruptions are prevalent. The situation represents a catastrophe waiting to happen and before long either a powerful eruption or major quake is going to score a direct hit on a poorly constructed megacity and its ill-prepared population. The outcome will be more horrific than any natural disaster experienced so far.

Over the past 1,000 years, earthquakes have caused nearly 8 million deaths. The ever-growing megacities, however, pose major new targets and it would be no surprise if, within the next few decades, 3 million urban dwellers were wiped out in a single massive quake. The threat from volcanic eruptions is barely less, and over 500 million people—1 in 12 of the Earth's population—now inhabit the danger zones around active volcanoes. Unlike earthquakes,

ABOVE As the Earth's population soars, so more people cram themselves into giant, urban centers. **BELOW** It may be only a matter of decades before a massive quake takes 3 million lives.

a major volcanic eruption can also impinge upon the population of the entire planet through its impact on the climate, and after-effects can exact severe retribution on harvests and health even on the other side of the world.

Even with business as usual, it seems that the toll from future tectonic hazards can only go up. The greatest fear of hazard scientists and disaster managers is that we will shortly be faced with a global natural catastrophe of unprecedented size. The threat of an asteroid or comet impact is now understood, but few appreciate that we are overdue for a volcanic super-eruption big enough to freeze the planet. Other tectonic threats include earthquake "storms," powerful enough to devastate a continent, and giant tsunami—huge sea waves normally generated by earthquakes—capable of obliterating entire cities around the Pacific and Atlantic Oceans.

HOW THE EARTH WORKS

The volcanic eruptions and earthquakes that make up the ever-present tectonic threat to our life on Earth are a constant reminder of the enormous energies generated in our planet's interior.

The Earth is an incredibly powerful heat engine driven by the breakdown of radioactive elements, which in turn drives geological processes that impinge upon all our lives. Our planet is in continual turmoil both below and at the surface—nothing is ever stable and nothing is entirely predictable. The crust and the uppermost part of the Earth's mantle, the layer immediately beneath the crust, form a rigid outer skin that behaves very differently from the rest of the planet. It is this outer skin that is broken up into huge rocky slabs, whose movements, relative to one another, are responsible for generating earthquakes and permit fresh magma to reach the surface at volcanoes. The boundaries of these giant tectonic plates form an interconnecting network across the face of the planet, marked in places by mountain ranges, huge tears in the continents, and, in other areas, by the great fissures and trenches that lurk deep beneath the oceans.

What happens in the Earth's rigid outer shell is simply a reflection of the dynamic situation below, and it is the slow churning of the mantle beneath that drives the plates in their slow crawl across the planet's surface. Ultimately, it is the Earth's metallic core that provides the enormous heat energy, first to the mantle, and then upward to the rigid exterior and the plates. Here, nearly 3,100 miles (5,000km) beneath the surface, is the Earth's heart, a part liquid, part solid ball of nickel and iron as hot as the surface of the Sun.

It is here that our planet's magnetic field is generated, without which the Earth and all life upon it would have far less protection from the potentially lethal cosmic rays of the Sun and deep space that constantly blast through our solar system. So, on the one hand, it is the core that drives the geological processes that make the Earth such a hazardous place; while, on the other, the same force makes the surface environment amenable to the evolution and survival of complex life forms such as ourselves.

OPPOSITE In Iceland, lava fountains mark the position of the Mid-Atlantic Ridge, where two tectonic plates are pushed apart. BELOW Earthquakes, landslides, and volcanic eruptions have killed perhaps ten million people over the last two millennia.

The world's greatest tectonic catastrophes

YEAR	COUNTRY	EVENT	FATALITIES
526	TURKEY	EARTHQUAKE	250,000
1290	CHINA	EARTHQUAKE	100,000
1303	CHINA	EARTHQUAKE	200,000
1556	CHINA	EARTHQUAKE AND LANDSLIDES	830,000
1622	CHINA	EARTHQUAKE	150,000
1731	CHINA	EARTHQUAKE	100,000
1815	INDONESIA	VOLCANIC ERUPTION/FOLLOWING FAMINE	90,000
1850	CHINA	EARTHQUAKE	300,000
1883	INDONESIA	VOLCANIC ERUPTION AND TSUNAMI	36,000
1902	MARTINIQUE	VOLCANIC ERUPTION	29,000
1923	JAPAN	EARTHQUAKE AND FIRE	142,800
1939	TURKEY	EARTHQUAKE	362,740
1970	PERU	EARTHQUAKE AND LANDSLIDE	67,000
1976	CHINA	EARTHQUAKE	290,000
1985	COLOMBIA	VOLCANIC ERUPTION TRIGGERED MUDFLOW	23,000
1999	VENEZUELA	LANDSLIDES	50,000
2001	INDIA	EARTHQUAKE	<100,000

Digging deep

>>>Like a rocky onion, the Earth is made up of concentric layers that reflect geological processes that operated early in its history.

>Over 4 billion years ago the Earth was a boiling, homogeneous mass of molten rock. Slowly, a range of geological processes conspired to separate out the different components and form the more organized interior of today.

The original crust formed slowly from the crystallization and cooling of magma from the Earth's interior but has since been strongly modified by the formation of sediments due to weathering and erosion, and by the metamorphic baking of both sediments and magmatic rocks deep within the crust itself. Although it presents us with a relatively stable environment on which to live, the crust makes up less than 1 percent of the Earth's volume and less than 0.5 percent of its mass.

Peeling away the layers

Very simply, the crust can be divided into two types, one flooring the ocean basins and the other forming the continental-land masses. Both types are derived from the crystallization of magma, with dark, dense basalts underlying the oceans, and paler, lighter granites forming the bulk of the continents. These have been much modified since their formation. Continental crust contains a plethora of other rocks such as limestones, sandstones, marbles, and clays, as well as exotic types like schists and gneisses.

We don't have to dig down too far to reach the base of the crust, which is a mere 6 miles (10km) thick beneath the ocean basins, although it may thicken to 50 miles (70km) beneath the highest mountain ranges. Here we enter the mantle, a part of the Earth about which little was known until the twentieth century, but which is now less enigmatic as a result of

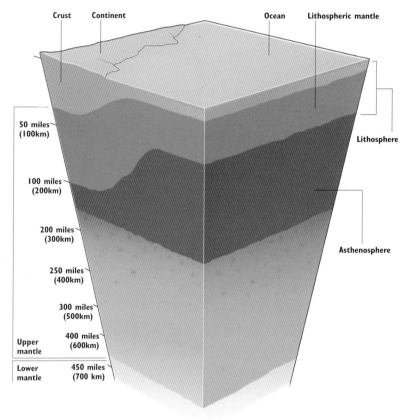

relatively new geophysical methods such as seismology. The mantle is huge, making up three-quarters of the planet's volume and two-thirds of its mass. For much of its bulk, it is solid, and although temperatures are extremely high—around 3,600°F (2,000°C) where it meets the core—the enormous pressures generally inhibit melting. The exception is a shallow layer of the uppermost mantle—often less than 100 miles (150km) deep—known as the asthenosphere, where the balance between pressure and temperature is just right to allow partial melting.

ABOVE The Earth's rocky plates are made up of rigid lithosphere that "floats" upon the partially molten asthenosphere.

From a hazard point of view, the asthenosphere is very important, being the source of all magma and, therefore, all volcanoes. Above this partially molten layer, the remaining overlying mantle and the crust are both solid and comparatively cold. Together they make up the lithosphere, the rigid outer layer of the Earth, which is broken up into a dozen huge plates and a number of smaller platelets. In broad terms the mantle comprises both upper and lower layers—compositionally very similar, consisting largely of the bright green mineral olivine and its various sister minerals. Olivine crystals sometimes reach the surface of volcanoes and are sought after by the jewelry trade, which markets them as the semiprecious stone peridot.

Once entirely off limits to our questing curiosity, the Earth's core has largely yielded up its secrets, substantially thanks to studies of the seismic waves produced during earthquakes that allow us to "see" deep into the Earth.

The boundary between the mantle and the core sits at just under a depth of 2,000 miles (3,000km) and, although the core comprises only 15 percent of the planet, its metallic nature ensures that it makes up about one-third of its mass. Like the mantle, the core consists of two layers, but here they are very different. While the inner core takes the form of a great sphere of solid iron and nickel, the outer core is a churning mass of liquid metal that generates the Earth's vital magnetic field and carries heat from the inner core to the lower mantle. From here it is transported onward to the upper mantle, where it drives the slow progression of the great plates across the Earth's surface.

BELOW A rare exposure of mantle rock in the Gros Morne National Park, Newfoundland, Canada.

Soup and cycles

>>>The moving plates that make up the Earth's outer layer consist not just of the crust, but of a composite of the crust and the thin, brittle, uppermost mantle.

>Together, the two constitute the lithosphere, which effectively "floats" on the underlying, partially molten mantle layer known as the asthenosphere. The concept of plate tectonics is crucial to our understanding of how the Earth works, and provides a guide to the nature and spatial distribution of tectonic hazards such as earthquakes and volcanic eruptions. But just what is plate tectonics? More specifically, just what forces are responsible for moving such enormous slabs of rock across the surface of the planet?

In order to understand plate tectonics we must return to the idea of the Earth as a gigantic heat engine, which is continually trying to cool itself down by transferring heat from the interior to the surface and from there into space. The ultimate driving force behind this tendency to lose heat is convection, a phenomenon that occurs on all scales from an entire planet to a pot of simmering soup. To examine the latter analogy more closely: heat is

transferred from a stove top to the base of a pot by a mechanism known as conduction, involving the transfer of heat energy from one atom to the next. When heat enters a fluid, however, the more efficient mechanism of convection takes over. Hotter liquid from the base of a soup pot rises to the surface in the

BELOW As two plates are pulled apart at a mid-ocean ridge, new magma rises to fill the gap.

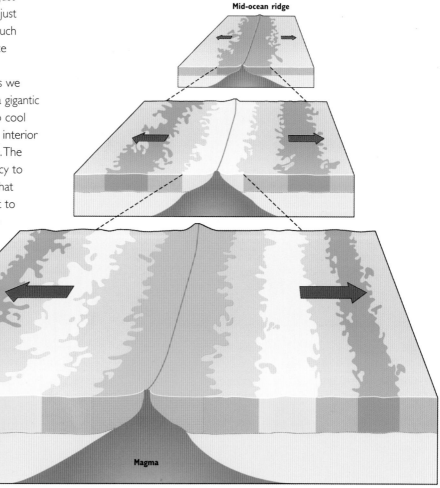

Mid-ocean ridge

Lithosphere　　　　Magma

center and moves out to the edge—cooling and becoming more dense as it goes—before sliding back beneath the surface.

In the beginning

The same process occurs within the Earth. Deeper, hotter, and, therefore, lower density material rises toward the surface, while the cooler—and more dense—material at higher levels in the mantle is displaced and sinks to take its place. A self-sustaining cycle is set up, involving the establishment of continuous currents known as convection cells, which constantly churn up the mantle while, at the same time, transferring heat outward from the core.

During its very early history, almost all the Earth's molten interior was involved in the convection process. The effect of the underlying convection cells on the thin, fragile crust was to break it into fragments and pull it apart where new hot material was rising from below, and to cause it to sink back into the inferno where two fragments collided. As the Earth cooled the primordial crust and uppermost mantle became more rigid and thickened to form the brittle lithosphere, and convection became confined to the remainder of the mantle. Today, rising convection currents bring hotter material from deep down to the relatively shallow depths of the asthenosphere. Here, the pressures exerted upon it fall, causing it to melt. At the same time the rising convection current reaches the base of the rigid lithosphere and starts to move laterally, dragging with it the rigid plate above.

Where two of the mantle's convection currents move laterally outward in opposite directions, the huge frictional forces generated tear the overlying rigid lithosphere apart, allowing new magma generated by melting in the asthenosphere to rise toward the surface. A constructive plate margin is thus formed, which repeatedly fills with magma as the two new plates are progressively pulled apart. After tens of millions of years, the two plates may be separated by thousands of miles. The magma that fills the gap between the plates is dense basalt, which forms low-lying topography that rapidly becomes occupied by the oceans. As a consequence, constructive plate margins are usually submarine, forming a global network of mid-ocean ridges tens of thousands of miles long. This ridge system is the birthplace of new basaltic oceanic lithosphere, which is constantly being created and forced away from constructive plate margins at a few inches a year—just about the same rate that fingernails grow.

BELOW At mid-ocean ridges, rising magma heats water in the rock, that rises in the form of mineral-charged hot springs known as black smokers.

Broken planet

>>> If new lithosphere is constantly being created at the mid-ocean ridge system, then, clearly, either the Earth is growing larger or a comparable amount of lithosphere is being gobbled up.

> As far as we know, the Earth has been pretty much the same size since its formation, so somewhere there have to be lithosphere eaters. Like the mid-ocean ridges, these are found beneath the oceans, in spectacular trenches, some deep enough to swallow Mount Everest. It is here that oceanic lithosphere, born at a mid-ocean ridge, ends its life. After a journey of thousands of miles, which may have lasted hundreds of millions of years, the basaltic oceanic lithosphere is cold, wet, and very dense. When it crashes into another plate moving in the opposite direction something has to give. Invariably, it is the coldest, densest plate that is forced to slide underneath, plunging down into the hot asthenosphere. The mechanism is known as subduction, and the deep trenches where this occurs are subduction zones. Not surprisingly, these graveyards of oceanic lithosphere are known as destructive plate margins.

As oceanic lithosphere is ultimately consumed at subduction zones, the floors of ocean basins are rarely older than a few hundred million years. The rocks of continents, in contrast, are often over 1 billion years old. The difference in age is because continents are not subducted. Instead of basalt, they are built from granite— the pale-colored, crystalline rock that faces many grand buildings. Because granite has a much lower density than basalt, when a continental plate collides head-on with an oceanic plate, it is the dense oceanic plate that loses the battle. Just like trying to push a large beach ball beneath the sea, subducting a continental plate is not possible because the granite plate is buoyed up by its low density. As a result, many continents are almost

as old as the Earth itself, growing incrementally in size over billions of years as volcanic eruptions and other geological processes add material to their margins.

Finding faults

Sometimes the Earth's plates scrape past one another instead of colliding head-on. These conservative plate margins mark some of the world's greatest geological faults and are home to some of the largest earthquakes ever known. Without doubt the best-known conservative plate margin is located in California, where the infamous San Andreas Fault marks the join between the Pacific Plate to the west and the North American Plate to the east. As the former grinds spasmodically northward at around 2 inches (5cm) a year, the region is shaken by destructive earthquakes that constantly threaten the cities of San Francisco and Los Angeles. Another conservative plate boundary is marked by Turkey's North Anatolian Fault, a particularly lethal earthquake generator that has taken thousands of lives over the centuries. The fault last jumped in 1999, flattening the city of Izmit and neighboring towns, and killing over 17,000.

If the faults that mark conservative plate margins moved smoothly past one another then dangerous earthquakes would not exist. Yet scraping two enormous rock masses past one another is like trying to slide two slabs of glue-smeared wood against each other. Nothing happens for a time, but as the strain increases and overcomes friction, so one slab of wood jerks forward. Plate boundary faults behave in

ABOVE The Earth's rigid exterior is broken up into a dozen gigantic rocky plates that move about the surface of the planet in the slow dance known as plate tectonics.

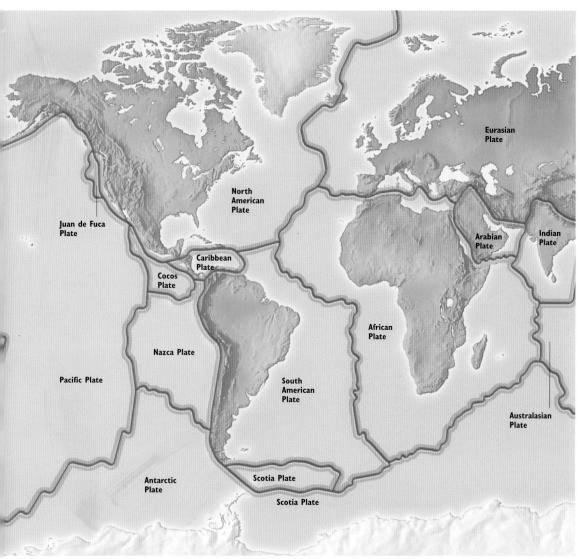

Juan de Fuca Plate

North American Plate

Eurasian Plate

Arabian Plate

Indian Plate

Caribbean Plate

Cocos Plate

African Plate

Nazca Plate

Pacific Plate

South American Plate

Australasian Plate

Antarctic Plate

Scotia Plate

Scotia Plate

BELOW New lithosphere is formed along the mid-ocean ridge system and destroyed in subduction zones where one plate is consumed beneath another.

just the same way. As the plate on one side of the fault tries to move against the other, huge strains accumulate that are eventually relieved by the fault jerking into life. When this happens, one of the plates is displaced laterally, releasing in a few seconds forces that may have been gathering for decades or centuries. The energies unleashed trigger the severe ground-shaking so characteristic of destructive earthquakes.

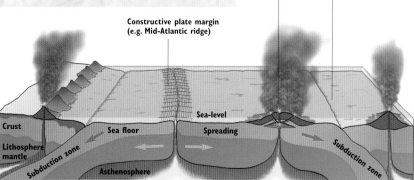

Mantle plume or "hot-spot" (e.g. Hawaii)

Destructive plate margin (e.g. Japan, western coast of South America)

Constructive plate margin (e.g. Mid-Atlantic ridge)

Crust

Sea-level

Lithosphere mantle

Sea floor

Spreading

Subduction zone

Asthenosphere

Subduction zone

THE DANGER FROM WITHIN

Luckily for some of us, volcanoes and large earthquakes are not distributed randomly across the entire planet. They are confined mainly to the active margins of the plates, rather than to the relatively stable interiors.

Most of the world's most violent volcanoes congregate along destructive plate margins and recent explosive eruptions at Mount St. Helens (U.S., 1980), Pinatubo (Philippines, 1991), and Soufrière Hills (Montserrat, 1995–present) all occurred at volcanoes sited above subduction zones. Strings of active volcanoes also lie along the mid-ocean ridge system, but these are usually lava-producing volcanoes and relatively nonexplosive. Barring Iceland and a few others, their eruptions are also mostly submarine and far from habitation. There are some exceptional intraplate volcanoes that hide in the hearts of the continents and ocean basins, where a plume of hot mantle material—known as a hotspot—forces fresh magma to the surface. Although often lava-producing and only mildly explosive, some hotspot volcanoes—such as those of Yellowstone, Wyoming (U.S.)—have blown themselves apart in some of the most violent volcanic blasts ever recorded.

Where volcanoes skulk, earthquakes are never far away, and indeed volcanic eruptions are almost always preceded and accompanied by quakes. Like volcanoes, most destructive quakes are confined to the margins of plates, either where two grind past one another, as at the San Andreas, or where one plate dives beneath another at a subduction zone. A glance at recent news reports will provide quite an accurate guide to those countries most susceptible —among others, Italy, Greece, Turkey, Iran, Afghanistan, India, China, Indonesia, Japan, Mexico, Nicaragua, Chile, and the United States. Plotting these on a map will reveal the two major earthquake belts, one circling the Pacific, and the other stretching from southern Europe through the Middle East and southern Asia, into China. Although earthquakes are also common along the mid-ocean ridge system, these are generally small and far from population centers. Large intraplate earthquakes are a threat, however, reflecting the release of strains that have been transferred incrementally from the margins to the hearts of the plates. Destructive intraplate quakes have been recorded in western Europe and along the east coast of the United States, so even here the danger from within may make itself known with little warning.

OPPOSITE Lava flows down the flanks of Java's Mount Merapi, one of the world's most active and damaging volcanoes. BELOW Survivors of the devastating 1999 Turkish earthquake pick over the rubble of their homes for possessions that might have withstood the terrible onslaught that took over 17,000 lives.

Where Vulcan reigns

>>>Vulcan, the mythological Roman god of fire, is always at work deep within the Earth, and the resulting magma spews forth on the planet.

BELOW The main types of volcanic eruption range from the quiet Icelandic to the violent Plinian.

Icelandic

Hawaiian

Strombolian

Vulcanian

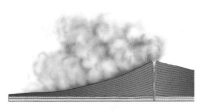

Peléan

>Even as I write this, the massive Popocatepetl volcano that towers over Mexico City has blasted more ash and debris into the skies above the city, while the Soufrière Hills volcano that has devastated much of the idyllic Caribbean island of Montserrat rumbles on into its seventh year. Historical records exist for eruptions at 539 volcanoes, but the true number of active volcanoes is probably much higher. Over 1,500 volcanoes have erupted since the Earth started to warm up again at the end of the last Ice Age around 10,000 years ago, and all are likely to erupt again. Some volcanoes, such as Stromboli and Etna (Italy) or Kilauea (Hawaii), are in almost constant eruption and in an average year about 50 volcanoes will burst into life. Although volcanoes are responsible for fewer deaths than earthquakes, they nevertheless inflict a terrible toll. Since the end of the eighteenth century alone, volcanic disasters have killed over 230,000

people and injured many thousands of others.

There is little doubt that the heart of Vulcan's kingdom lies in the Pacific Ocean, encircled by a "ring of fire" comprising over 1,000 active and potentially active subduction zone volcanoes. These particularly violent and destructive volcanoes are fed by magmas generated by the melting of oceanic crust as it plunges ever deeper into the furnace of the upper mantle. Because the subducting crust is particularly soggy after its interminable journey from a mid-ocean ridge, the magmas are also water rich. This unstable water/magma mix is partly responsible for the violent activity of these subduction zone volcanoes, but another cause is the magma composition, which is notably high in silica, a combination of oxygen and silicon atoms that binds the magma together and makes it sticky. This high viscosity in turn traps in gases, including water, until the magma reaches the low

Plinian

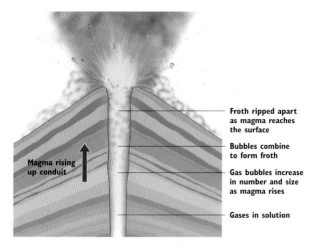

Froth ripped apart as magma reaches the surface

Bubbles combine to form froth

Gas bubbles increase in number and size as magma rises

Gases in solution

Magma rising up conduit

ABOVE As viscous magma reaches the surface, it is torn apart by expanding gas bubbles, resulting in a violent explosion.

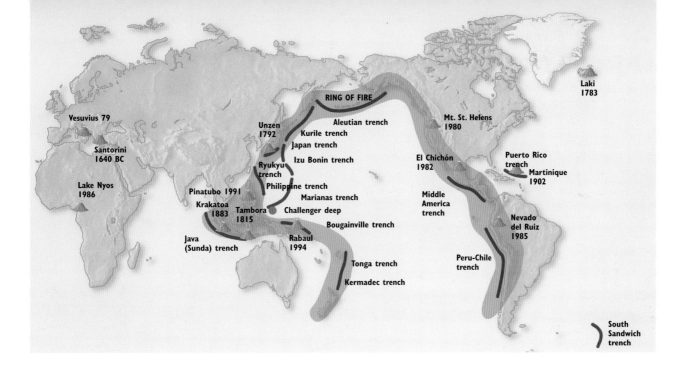

ABOVE Every year around 50 volcanoes erupt, many in the Ring of Fire that circles the Pacific Ocean.

pressures at the surface, when the gases rip the molten rock apart in a devastating blast.

Many of the most violent eruptions of recent centuries have been recorded in the southeast quadrant of the Pacific, particularly in Indonesia. Here, the cataclysmic explosion of Tambora in 1815 led to a drastic cooling of the northern hemisphere's climate, while Krakatoa generated the loudest noise ever heard when it blew itself to pieces in 1883. Across the Pacific a much less violent—but no less significant—volcanic event took place in a farmer's field, when within a few months a smoking hole transformed itself into the volcano now known as Paricutin.

Trouble in paradise

Vulcan's violent warriors also terrorize other parts of the world where one plate dives beneath another, most notably in the Caribbean —host to the greatest volcanic disaster of the twentieth century. Here, the islands of the Lesser Antilles provide a base for 17 active volcanoes, of which, on average, one erupts every 12 years. In 1902 the town of St. Pierre on the island of Martinique was obliterated by a terrible eruption of Mont Pelée, which—after 50 years of slumber—unleashed a hurricane-force blast of magma fragments, hot ash, and incandescent gases straight at the town. Of around 30,000 inhabitants, only two horribly burned survivors crawled away from the incinerated rubble.

Away from subduction zones, Vulcan's soldiers adopt a more benign attitude toward the local inhabitants. At Kilauea and Mauna Loa in Hawaii, for example, and at Iceland's many volcanoes, the main product of eruptions is fast-flowing lava, which, although highly effective at destroying property and inundating farmland, rarely takes life. Such volcanoes are characterized by magma that is much runnier and less gas-rich than that produced by their subduction zone counterparts, resulting in eruptions that are considerably less violent. Even here, though, the occasional explosive eruption is possible, and the violent blast at Askja (Iceland, 1875) sits in the top 20 of the largest explosive eruptions of the nineteenth and twentieth centuries.

Quake country

>>>Our restless planet is almost constantly shaking, and over half a million quakes are measured every year—an astonishing 1 per minute.

>Fortunately for us, over 99 percent of these are too small to be damaging and most are so weak that they can only be detected by sensitive seismographs—instruments designed to detect even the tiniest tremblings of the Earth. More than half of the remaining 1 percent of quakes are large enough to cause concern and damage, and about 9 of these occur every day, almost anywhere on the planet. Usually, the resulting damage is minimal, limited to smashed dishes, broken windows, and cracked walls—serious injury or death is rare. Severe problems don't really start until quakes reach magnitude 6 on the Richter Scale, but such large events are normally confined to plate margins. A quake on this scale shakes the planet about every 3 days and can cause extensive damage and loss of life. Its impact, however, depends on which part of the world it hits.

Buildings kill, not earthquakes

In California, where strict construction codes are enforced to ensure that buildings can withstand even severe earthquakes, a magnitude 6 event would cause little damage and probably no loss of life. The situation would, however, be very different if the same size quake hit a poorly constructed city in a developing country where such codes do not exist or are not enforced. Early in 1999, for example, an earthquake registering magnitude 6 struck a mountainous region of western Colombia, totally flattening several towns including the city of Armenia, which was virtually reduced to rubble. The

resulting death toll was over 2,000 with many thousands of others injured or homeless. Although the size of an earthquake is directly related to its impact, this is clearly not the only factor that determines the level of death and destruction. As earthquake engineers are always keen to stress, "it is buildings, not earthquakes, that kill people." As poorly planned and badly built towns and cities continue to expand in earthquake-prone regions of the developing world, the effects of even moderate quakes will inevitably become increasingly disastrous.

The destructive capacity of an earthquake is also dependent on a range of geological factors. For example, the shallower an earthquake and the longer its duration, the more potentially damaging it is. The nature of the underlying rock is also critical. Buildings constructed on solid bedrock will normally fare well, even during a large quake, while those on soft sediment, landfill, or reclaimed land will almost certainly not survive intact. This is because the severe shaking of the ground causes the sediment to liquefy

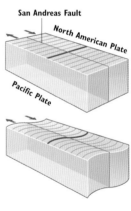

San Andreas Fault
North American Plate
Pacific Plate

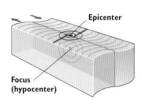

Epicenter
Focus (hypocenter)

OPPOSITE TOP No place is entirely free from earthquakes. In 1884, a quake in southeast England damaged over 1,000 buildings.
OPPOSITE BELOW An earthquake forms when strains accumulating in two huge masses of rock are instantaneously released, causing the deformed rocks to rebound to their initial strain state, like an elastic band after is has snapped.
RIGHT The locations of over 30,000 deep quakes recorded over a five-year period clearly mark out the positions of the plate margins.
BELOW The enormous stresses acting close to California's San Andreas Fault cause the rocks to buckle and fold.

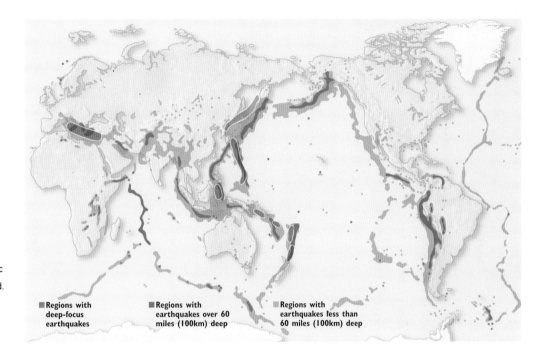

■ Regions with deep-focus earthquakes

■ Regions with earthquakes over 60 miles (100km) deep

■ Regions with earthquakes less than 60 miles (100km) deep

and act like a fluid. Anything heavy sitting on the surface—such as a house—will simply sink.

Large earthquakes are not, however, just about collapsing and sinking buildings. They also trigger a whole range of terribly destructive phenomena such as landslides, fires, and the giant sea waves known as tsunami. Most lethal landslides are due to the same severe ground-shaking that brings down buildings, and are particularly serious in steep, mountainous terrain. In 1970, a big quake in Peru brought down an entire ice-covered mountain peak that, within minutes, obliterated several towns. Tsunamis are also a widespread consequence of earthquakes, particularly when they happen beneath the seabed. Traveling at the speed of a jet aircraft, quake-generated tsunami can travel across an entire ocean within 24 hours, bringing death to unsuspecting settlements thousands of miles from the earthquake source.

Of all the appalling consequences of a major earthquake, however, fire is probably the most feared. Exploding fuel tanks, fractured gas mains, and overturned stoves all conspire to create a conflagration once the ground has stopped shaking, and the results can be horrific. Gigantic firestorms, whipped up by strong winds, raged for days after the 1923 Tokyo quake, roasting tens of thousands alive as they tried to outrun or shelter themselves from the flames.

LESSONS FROM HISTORY

The threat from earthquakes and volcanic eruptions has always been with us and, indeed, evidence from our ancient past shows that our distant ancestors had an even tougher time.

It is hard to imagine that barely 10,000 years ago our planet was still in the fierce grip of the last Ice Age, with much of the land under a blanket of ice and ocean levels dramatically lower. As the planet started to warm and the ice melted, so all sorts of dynamic changes began to take place that contributed to an increased tectonic threat to budding civilizations. Rising sea levels promoted instabilities along coastal margins, causing long quiescent volcanoes to spring into life and previously subdued faults to trigger shattering earthquakes. As the oceans rose rapidly, so gigantic landslides poured off the edges of the continents, generating huge tsunami that sped far and wide. Around 7,000 years ago, a pile of sediment the size of Wales collapsed into the North Atlantic from the Norwegian continental shelf, sending waves 50 feet (15m) high crashing into northeast Scotland.

At around that time, the rising Mediterranean finally smashed its way through a zone of crustal weakness, sending a raging torrent of water into the lower-lying Black Sea via the greatest waterfall ever known, and forcing the mass migration of those peoples living around what was then just a large freshwater lake.

As the Earth's crust continued to strain and creak beneath the rapidly expanding oceans, huge volcanic explosions choked the skies. Around 5,000 years ago, a great mass of rock from Mount Etna plummeted into the sea, perhaps prompting the inhabitants of eastern

Sicily to head further west. Just 3,600 years ago the demise of the great Bronze Age civilization of the Minoans was heralded by the obliteration of the Greek island of Thera in a terrible volcanic cataclysm that spawned the Atlantis myth. The eruption may even have been violent enough to part the waters of the Red Sea and allow the fleeing Israelites passage.

The carnage continued into the Christian Era, with Pompeii wiped out by a rejuvenated Vesuvius in 79 A.D. and the whole of eastern Europe shaken by huge quakes during early Medieval times. However great the tectonic threat today, the lesson from history is clear —things can get much, much worse.

OPPOSITE The parting of the Red Sea may have been the waters withdrawing prior to the arrival of a deadly tsunami triggered by the volcanic obliteration of Thera. BELOW Recent submarine surveys have revealed signs of a long-drowned civilization below the waves of the Black Sea.

Myths and legends

>>> We have to turn detective in order to learn more about the natural catastrophes of the past.

> Ancient texts such as the Bible are packed full of dramatic and terrifying accounts of mass destruction and enormous loss of life, few of which have been taken seriously by scientists and archaeologists until recent decades. Increasingly, however, more careful reading between the lines has revealed that many of these stories can be interpreted as somewhat embellished accounts of the wrath of Mother Nature rather than of a vengeful deity.

Fact or fiction?

The catastrophe-related myths that have attracted most attention are those which are particularly persistent, such as the biblical Great Flood charged with wiping out most of humanity. This story can be traced back to the Babylonian account of Gilgamesh—ruler of ancient Sumeria *ca.* 2600 B.C.—which invokes a similar deluge that decimates the human population. This tale in turn appears to have its roots in the Atrahasis legend, an earlier Sumerian account that talks of a terrible flood that must have occurred before 3000 B.C. Could the truth of the matter be that these accounts reflect an ancient folk memory of the flooding of the Black Sea region around 7,000 years ago? Some archaeologists and scientists think so, suggesting that some of the tribes forced out by the rapidly rising waters made their way southeastward, eventually settling in Sumeria where their unlikely tale found its way into local legend.

OPPOSITE TOP LEFT According to the Lamas of Mongolia, earthquakes resulted from the frog on which the Earth rested lifting one of its legs.

BACKGROUND Could the biblical Great Flood be based upon an ancient folk memory of the flooding of the Black Sea? **OPPOSITE RIGHT** Native Americans thought that the striations on the Devil's Tower in Wyoming had been gouged out by a giant bear.

Ancient texts also shed light on our ancestors' interpretations of the tectonic hazards that faced them, which invariably involve the retribution of an irate deity rather than an exploding volcano or shuddering fault. Volcanoes have been held in mystical awe since time immemorial and they appear in cave paintings as far back as 6000 B.C. Volcanoes and the gods have always existed side by side, and both Hephaestus—the Greek god of fire—and his Roman equivalent, Vulcan, plied their smithing and smelting trades deep beneath the roots of active volcanoes. Other societies found equally extraordinary explanations for volcanoes and volcanic features. The Native Americans, for example, blamed the striations on the sides of the 985 feet (300m) high volcanic plug known as Devil's Tower on the claws of a giant, supernatural bear. To the Aztecs, the fires from their volcano came from the god Popocatepetl, who stood on the summit holding an eternal funeral torch for his lost love Iztaccihuati. The volcanoes of Hawaii were believed to be manifestations of the fire goddess—Pele—who, according to legend, was chased by her sister from west to east along the island chain, temporarily seeking shelter in each volcano in turn before being discovered and moving on to her current home on the easternmost island of Hawaii. The amazing thing about this tale is that the route followed by Pele

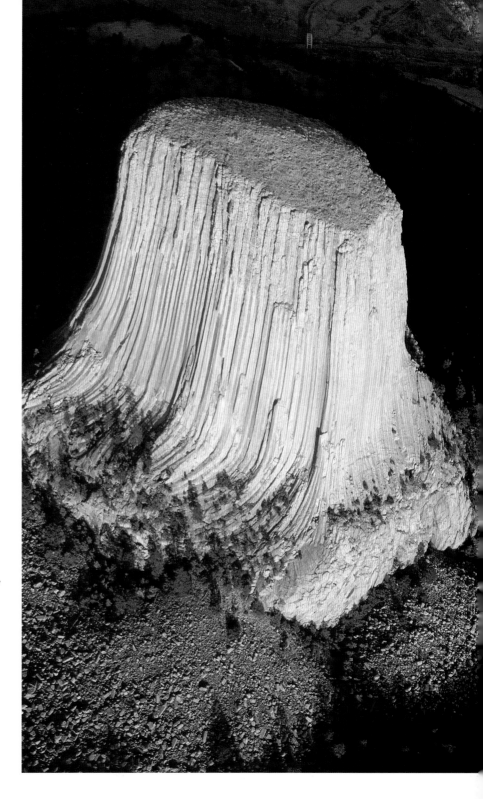

does actually parallel the sequence of volcanism in the island chain.

Myths and legends—both written and oral—are packed with tumbling towns and collapsing cities, and it is tantalizing to examine these doom-laden accounts in the context of large earthquakes. Certainly, many of the biblical tales of fallen cities are likely to have been based upon the real-life destruction of major urban centers of the ancient world. The Middle East is a major earthquake zone and the Jordan Rift that slices north–south through the Holy Land, and its associated faults, have in the past generated severe quakes. The walls of Jericho were almost certainly brought tumbling down by a large earthquake, while another is likely to have been responsible for obliterating the fleshpots of Sodom and Gomorrah. In the Old Testament, earthquakes are used by God to punish humans for their sins. Other societies, however, had rather different ideas. According to the Lamas of Mongolia, the Earth shook when the giant frog on which it rested lifted a leg, while the Timorese thought earthquakes were caused when the giant supporting our planet switched it from one shoulder to the other.

The end of Atlantis?

>>>The legend of Atlantis is perhaps the greatest and most enduring myth of western civilization; but does it have a root in fact?

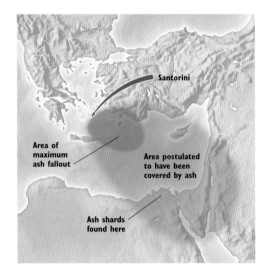

Santorini

Area of maximum ash fallout

Area postulated to have been covered by ash

Ash shards found here

with contributing to the fall of the great Minoan civilization, which, although centered on Crete, also had settlements on Thera.

For a people that walked the Earth 1,600 years before Christ, the Minoans' quality of life was quite astounding. Excavations on Thera have exposed a standard of living better than that in many parts of the world today, with three- and four-story stone houses containing running water and toilet facilities. Skillfully wrought tools, pottery, and glassware have been excavated in abundance, and spectacular frescoes have been revealed that show men and women as equals and are unique in the ancient world in their absence of militaristic imagery. At the time, the

FAR LEFT Can the volcanic obliteration of Thera provide an explanation for the Old Testament accounts of the Egyptian plagues? BELOW The Atlantis legend may have its roots in the terrible destruction of the Minoan island of Thera over 3,500 years ago.

>According to the fourth century B.C. Greek philosopher, Plato, Atlantis was an ancient seafaring nation located beyond the Gibraltar Straits in the Atlantic Ocean, which sank beneath the waves in a gigantic cataclysm. Despite detailed surveys of the Atlantic seafloor no trace of such a civilization has been found and most geologists and archaeologists now place Atlantis much closer to the heart of the Mediterranean.

Tree-ring analysis shows that the year 1626 B.C. was much colder than normal, hindering plant growth in Europe and North America and resulting in the formation of thinner rings of new wood. The unusually cold conditions were not simply the result of our planet's capricious climate, but were caused by a gargantuan volcanic explosion that shattered the Greek island of Thera—now more commonly known as Santorini. The huge blast—one of the most violent ever recorded in the Mediterranean—is charged by some scientists and archaeologists

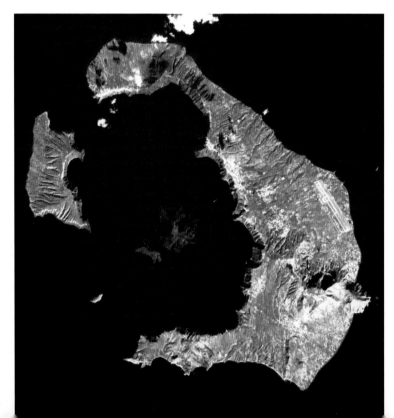

Minoan culture must have had an extraordinary influence across the eastern Mediterranean and perhaps far beyond. Indeed, to those who toiled and suffered in poorer and less sophisticated societies, it might quite easily have been elevated to mythical status even while still in existence. It is not difficult to imagine that its sudden demise led, over succeeding centuries, to the legend of a mystical nation of great richness and culture that vanished beneath the oceans.

A civilization blown apart

The Thera volcano must have been rumbling for weeks or even longer before the end, and the absence of human remains at Akrotiri suggests that this part of the island at least had already been evacuated. The climactic phase dumped 65 feet (20m) of ash and debris across the island, before seawater mixing with magma literally blew the volcano apart. This would have obliterated all traces of the Minoans from Thera; but could the eruption have struck at the Cretan heart of the Minoan civilization?

Bearing in mind that Crete is just a day's boat journey from Thera, many think that this is quite likely. Geological studies supported by computer modeling suggest that the earth-shattering blast generated tsunami that sped out across the entire eastern Mediterranean. It now seems likely that the coast of northern Crete, which hosted most of the Minoan settlements, was scoured by huge waves up to 330 feet (100m) high, dealing a devastating blow to the seafaring nation. Down, but not quite out, the Minoans continued to trade after the cataclysm, but in just a few generations their now struggling society was invaded and replaced by Mycenaeans from the Greek mainland.

The destruction of Thera may provide an explanation not just for the legend of Atlantis, but also for the Old Testament accounts of the Egyptian plagues, and the timing is, in fact, just

about right. Some have suggested that the accounts in Exodus of three days of darkness might reflect heavy ash fall, while the "river of blood" might be the massive rafts of pink Theran pumice. Even more intriguing, receding waters from the marshy northern end of the Red Sea might have offered the Israelites an escape route before the tsunami rushed back to drown the pursuing Egyptian soldiers.

ABOVE The Minoans' quality of life exceeded that in many parts of the world today.
BELOW The legacy of the titanic volcano blast is an enormous sea-filled crater bounded by spectacular cliffs.

Paroxysms of the past

>>> There is growing evidence to suggest that even the worst earthquakes of recent years pale in significance when compared with the great, earth-shattering quakes of the past.

ABOVE Did Troy really burn as a result of conflict and war or was a massive earthquake to blame?

> Instead of single events, it seems that some periods in our relatively recent history were characterized by clusters of destructive earthquakes—sometimes called earthquake storms—that, within a few decades, devastated huge areas and obliterated many cities. The earliest of these storms may have battered much of the eastern Mediterranean and the Aegean during the Late Bronze Age, where archaeological evidence supports the fall of city after city across the region. Until recently, war and invasion were widely held responsible for the destruction of cities as far apart as Mycenae in Greece, Troy in Turkey, Aleppo in Syria, and Bethel in what is now Israel, between 1225 and 1175 B.C. Now, however, it seems that this unprecedented 50 year long episode of destruction may have resulted from a series of huge earthquakes occurring on the extensive network of active faults that cross the region.

The walls came tumbling down

This reinterpretation has been put forward by U.S. researchers Amos Nur and Eric Cline following a careful analysis of reports from excavations at nearly 50 cities in Greece, Crete, Cyprus, Turkey, and the Middle East. Many of these describe damage that can be more easily explained by an earthquake than by human activities, including the presence of patched and reinforced walls, crushed skeletons lying beneath debris, slipped keystones in arches and doorways, and rows of columns that have all collapsed in the same direction.

One of the cities believed to have been struck by a huge quake is the legendary city of Troy in eastern Turkey. Troy was destroyed in 1250 B.C. by some event that brought down and cracked many of the city walls and left piles of debris across the city. Given that Troy is not far from the North Anatolian Fault, whose rupture generated the huge quake that killed over 17,000 people in western Turkey in 1999, it should not be at all surprising if this ancient city also succumbed to movement of the fault.

The principle underlying the earthquake storm concept is known as stress transfer. If part of a fault system moves to trigger an earthquake, the result of the movement can be to transfer stress to another part of the fault, which will in turn move to trigger a second quake. This is, in fact, the mechanism by which earthquakes have marched westward along Turkey's North Anatolian Fault over the past 60 years or so; the next stop being Istanbul. In the Late Bronze Age, one large quake toward the end of the thirteenth century B.C. may have changed the stress pattern of the fault systems of the eastern Mediterranean in such a way as to trigger a cascade of further large quakes that felled some of the great cities of the time.

Remarkable though this may seem, the event may not be unique, and a similar earthquake storm seems to have shaken the region from the middle of the fourth to the middle of the sixth century A.D. Then, the eastern Mediterranean was hit by a number of great quakes that conspired to thrust skyward by up to 30 feet (9m) a gigantic block of the Earth's crust the size of Turkey. This Early Byzantine Tectonic Paroxysm appears to have devastated many communities over an enormous area, and accounts from surviving texts report great destruction and heavy casualties due both to quakes and to resulting tsunami. The peak of seismic activity seems to have been reached in 365 A.D. when earthquakes are reported to have struck Cyprus, Crete, and Libya, generating devastating tsunami all around the eastern Mediterranean coastline.

BELOW The Late Bronze Age cities of the eastern Mediterranean may have been toppled one after the other by a great earthquake storm.

THE NEXT THROW OF THE DICE

Disaster scientists are bombarded time and time again with questions containing the "I" word—"If" an asteroid hits the planet how many people will die?

If there is a super-eruption in Asia, will I be affected in New York? If there is another earthquake in Los Angeles, where is the safest place to seek shelter? Implicit in the use of the word "if" is that there is at least some chance that these events might never happen again; as if, somehow, the Earth might stop functioning as it has done for 4.6 billion years and, for no apparent reason, suddenly become a safer and friendlier place.

The whole problem comes down to personal experience. Even in a region that has already been struck by major earthquakes or volcanic eruptions, if things have remained quiet for two or three generations, then the inhabitants simply ignore the threat. This attitude is not necessarily conscious and intentional, but is more a combination of ignorance, lack of concern, and a general feeling that perhaps it won't happen to them. The population of Memphis, Tennessee, for example, are constantly made aware that their city will be struck sometime in the future by a large earthquake, but somehow each person feels that it will not be in their lifetime. And do the inhabitants of the New Zealand city of Auckland—surrounded on every side by cinder cones—ever pay more than lip service to the volcanic threat within their midst?

Such a lackadaisical attitude makes effective disaster preparedness extremely difficult, even in regions where tectonic hazards are common. For global catastrophes that only occur every tens or

hundreds of thousands of years it is almost impossible to galvanize interest and concern, even from governments and fellow scientists. It is vital, however, to do so, because the Earth remains just as dangerous as it always has been. The "I" word simply does not apply, but the "W" word does. The question we should be asking ourselves is—"When" will our society have to face a global tectonic catastrophe? The dice has already been thrown and is currently bouncing and bobbling across the board. When it finally comes to rest will it read "earthquake storm," "super-eruption," or "giant tsunami"? Only time will tell.

OPPOSITE A peaceful life on Graciosa in the Azores is sometimes rudely interrupted by a new volcanic eruption. Persuading the population that the threat is always present is a problem. BELOW Of the 16 largest eruptions of the last 200 years, 12 had never erupted in historical times.

Largest eruptions of the last 200 years

YEAR	VOLCANO	FIRST HISTORICAL ERUPTION?	DEATHS
1991	CERRO HUDSON (CHILE)	NO	0
1991	PINATUBO (PHILIPPINES)	YES	800
1982	EL CHICHÓN (MEXICO)	YES	2,000
1980	MOUNT ST. HELENS (U.S.)	NO	57
1956	BEZYMIANNY (KAMCHATKA, RUSSIA)	YES	0
1932	CERRO AZUL (CHILE)	NO	0
1912	NOVARUPTA (ALASKA, U.S.)	YES	2
1907	KSUDUCH (KAMCHATKA, RUSSIA)	YES	0
1902	SANTA MARIA (GUATAMALA)	YES	>2,000
1886	TARAWERA (NEW ZEALAND)	YES	>150
1883	KRAKATOA (INDONESIA)	NO	36,417
1875	ASKJA (ICELAND)	YES	0
1854	SHEVELUCH (KAMCHATKA, RUSSIA)	YES	0
1835	COSIGUINA (NICARAGUA)	YES	5–10
1822	GALUNGGUNG (INDONESIA)	YES	4,011
1815	TAMBORA (INDONESIA)	YES	92,000

Statistics—and more statistics

>>> Crucial to understanding the risk posed to our planet by global tectonic hazards is an appreciation of the concept of an event being "overdue."

> We are familiar with this term mainly from waiting in vain for buses or trains scheduled to arrive 15 minutes earlier. The Earth, however, does not operate like clockwork and as a result tectonic hazards do not run on a timetable. So they cannot, in truth, ever be said to be overdue in the same way as a late bus or train. In fact, events such as earthquakes and volcanic eruptions actually occur randomly in time but with a characteristic frequency known as the return period. Take, for example, volcanic super-eruptions that have plunged the world into the depths of volcanic winter, on average, every 50,000 years or so. The last gargantuan blast occurred at Toba, Indonesia, 74,000 years ago, but this does not mean that we are 24,000 years overdue for the next one. The Earth just doesn't work that way. Our planet is a highly complex system within which all the conditions needed for a super-eruption come together on average 50,000. In theory, therefore, it is possible—though unlikely—that we will have to wait 500,000 years for the next great eruption; or that two will occur next month.

Having said this, most people in the southern U.K. still remember that the devastating one-in-two-hundred-year "hurricane" that struck in 1987 was followed by another similar size storm just three years later. Furthermore, 1991 saw two major (though far from super-eruption status) volcanic eruptions when only one is normally expected every century.

These last two examples also teach us an important lesson about statistics. The fact that a rare event has just happened does not provide protection against its recurrence in the short term. Remember, even after throwing a six, the chances of another six on the second throw remain one in six—no more, no less.

Probabilities of death

Just what then are the average return periods of global tectonic hazards? Fortunately for us they are pretty low. This is because all tectonic hazards follow what is known as a power-law distribution, which means—technically speaking—that the number of any size of event is proportional to the inverse of the square of its size. Put simply, this means that there are many more small earthquakes and volcanic eruptions than large ones. From this it follows logically that the return periods of small tectonic hazards are very short and become progressively longer as the events increase in size. Thus, the average return period of magnitude 3 earthquakes, of which hundreds of thousands occur every year, is just five minutes, while that for the magnitude 8 city-levelers is a full six months. Similarly, volcanic blasts on the scale of Mount St. Helens are relatively common, causing regional devastation

BELOW Like buses, tectonic hazards run on a timetable, known as a "return period." Also, like buses, this can rarely be relied upon.

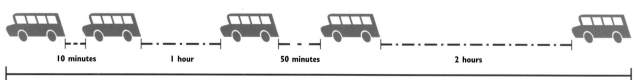

| 10 minutes | 1 hour | 50 minutes | 2 hours |

4 hours (average return period 1 hour)

about once a decade, while eruptions as big as Tambora in Indonesia, held responsible for the awful northern hemisphere Summer of 1816, only appear on the scene every millennium or so.

Perhaps the most extraordinary tectonic hazard statistic relates to the chances of anyone dying today in a global natural catastrophe. Figures for the United States, for example, reveal probabilities of one in 300 that a citizen will be murdered, one in 5,000 that they will be electrocuted, and one in 10,000 that they will die in an air crash. Quite astonishingly, the probability that a U.S. citizen will die as a result of a super-eruption could be as high as one in 5,000, making it twice as likely as being killed in an air crash. While it might sound ludicrous, it is based upon estimates that the freezing volcanic winter that follows a super-eruption will kill around a billion people—mainly through starvation and disease. If we could travel a million years into the future and then look back, we would see that in the intervening period perhaps 20 such gigantic blasts would have taken twice as many lives as crashing aircraft. A sobering statistic if ever there was one.

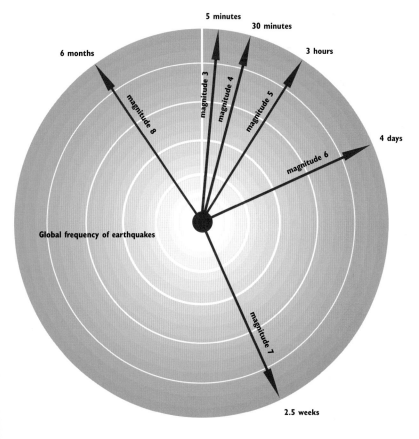

Global frequency of earthquakes

5 minutes · 30 minutes · 3 hours · 4 days · 2.5 weeks · 6 months

magnitude 3 · magnitude 4 · magnitude 5 · magnitude 6 · magnitude 7 · magnitude 8

Eruption numbers versus size

100,000
10,000
1,000
100
10
1
0.1
0.01

Number of eruptions per thousand years

2   3   4   5   6   7   8

Increasing size of eruption as shown by the Volcanic Explosivity Index (VEI)

LEFT Fortunately for us, there are much fewer large volcanic eruptions than small ones. Super-eruptions occur perhaps every 100,000 years.

ABOVE While the Earth trembles every five minutes or so, a major earthquake strikes only a couple of times a year.

VOLCANOES

VOLCANIC BLASTS FROM THE PAST

Volcanic disasters are nothing new. Ever since our nomadic ancestors at last settled down and began to build permanent settlements, our society has been periodically battered by volcanic eruptions.

Wherever the distinctive smoking cone appears, so will human beings—attracted by the fertile volcanic soils and, in the tropics at least, by the gentler climate the higher altitudes of a large volcano offer. Those who work the land around active volcanoes are aware of their Jekyll-and-Hyde characteristics and invariably, while accepting the fruits of Vulcan, they keep a constant eye on the smoking summit. In past times, however, the threat from volcanoes was often not realized and, if no signs of life were evident, an active volcanic cone was not distinguished from any other mountain. Without widespread use of the written word, memories were short, and within a few generations recollections of past eruptions and past catastrophes were quickly lost. If a list of active volcanoes had been put together at the time of Christ, it would have contained only a handful of Mediterranean volcanoes such as Etna and Stromboli. A list recently compiled by the Smithsonian Institution in Washington contains over 550 volcanoes at which historical eruptions have been recorded, and another 1,000 or so that are still clearly "alive."

Before the twentieth century, the first time a volcano was recognized as such by its local population was when it erupted, usually after a long period of repose—and this was certainly the case for the reawakening of Vesuvius in 79 A.D. Realization came too late, and at least 3,500 inhabitants of Pompeii and Herculaneum

succumbed to the devastating pyroclastic flows and surges that poured from the mountain. Since then, people all over the planet have repeatedly been taken by surprise by volcanic blasts, from freezing Iceland to sweltering Indonesia. In all, volcanoes are estimated to have killed 310,000 people before the dawn of the twentieth century, some in titanic explosions such as those of Vesuvius or Krakatoa, but many others in smaller attritional events. Volcanoes and humankind have fought a constant war throughout history and there is no doubt that the battle will continue.

OPPOSITE
The 2001 eruption of Mount Etna on the island of Sicily caused little damage, but an eruption in 1669 destroyed much of the nearby city of Catania.
BELOW After a long period of repose. Mount Vesuvius exploded into life in 79 A.D., wiping out the towns of Pompeii and Herculaneum.

The petrified city

>>> Life could rarely have seemed so good as for the inhabitants of Pompeii and Herculaneum in the opening century of the first millennium A.D.

> Nestled on the coast of the Bay of Naples, between Vesuvius and the warm waters of the Mediterranean, the twin Roman towns thrived both culturally and economically. Life was sweet, with appetites satisfied by the fruits of the fertile volcanic soils—a legacy of earlier eruptions of Vesuvius—that covered Mount Vesuvius's flanks.

Although a number of natural philosophers had wondered about the distinctive conical shape of Vesuvius, few in Herculaneum and Pompeii gave this a thought, and any suggestion that the local mountain was a dormant volcano would have been treated with derision. It seemed that little could disturb the wealth and tranquility of the region—until the peace was shattered by a violent earthquake in 62 A.D. Both towns were severely damaged by the quake, but rebuilding soon started, and even more impressive buildings began to emerge from the rubble of the old. It looked as if Pompeii and Herculaneum would be booming centers again.

Pressure cooker

Nature, however, had other plans. As well as devastating the surface, the quake had also shaken up the plumbing system beneath Vesuvius, allowing fresh magma to rise and pond within the volcano. Fifteen years later, the magma simmered like a pressure cooker, causing the volcano to bulge upward and generate increasingly severe earthquakes that shook the region. Close to midday on August 24, 79 A.D., the pressures could no longer be contained and the cooker exploded in a titanic blast that sent a column of pumice and ash 12 miles (20km) into the sky. The prevailing winds meant that Pompeii

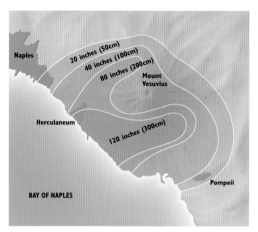

ABOVE The 79 A.D. eruption of Mount Vesuvius buried the nearby towns of Pompeii and Herculaneum in up to 10 ft (3m) of volcanic debris.

—just 5 miles (8km) southeast of the summit— took the full brunt of the volcano's fury during the early stages of the eruption. A rain of sharp rocks, ash, and hot cinders battered the town and its inhabitants, provoking panic and mass evacuation. Some decided to stay, sure that they were safe in their homes, but they were wrong—much worse was to come.

Early in the morning of August 25, Vesuvius directed the first of several pyroclastic surges at the town—hurricane blasts of superheated gas, pumice, and ash, against which there was little protection. In the open, the surge caught many who had returned to the town to collect valuables, incinerating them instantly in a wave of blistering heat. Most died before they

had time to think, their airways and lungs fried and the muscles of their limbs contorted into pugilistic poses.

Vesuvius then turned on Herculaneum, a mere 4 miles (7km) to the west. Just after midnight the huge column of debris that Vesuvius continued to discharge into the atmosphere could no longer sustain its own weight. It collapsed back on itself, shooting a terrible ground-hugging blast of incinerating gases, ash, rocks, and steam straight at the town. Within five minutes, the blast flattened buildings, roasting alive any inhabitants who remained. From the safety of Cape Miseno, 12 miles (20km) away across the bay, Pliny the Younger described the event:

"…on the other side, a black and dreadful cloud bursting out in gusts of igneous, serpentine vapor now and again yawned open to reveal long, fantastic flames, resembling flashes of lightning but much larger…"

The extraordinary feature of the 79 A.D. eruption of Vesuvius is that excavations at both Pompeii and Herculaneum have ensured that we can still see many of their inhabitants in their death throes—preserved forever as plaster casts made of the spaces amid the ash that their long-gone bodies once occupied.

LEFT Plaster casts, made from the spaces left by the rotting bodies encased in the ash, capture the awful death throes of the unfortunate inhabitants of Pompeii.

ABOVE Pompeii and Herculaneum were obliterated and buried by pyroclastic surges— incinerating hurricanes of superheated gas, pumice, and ash.

RIGHT The excavated ruins of Pompeii reveal a bustling and prosperous Roman metropolis that was wiped out in minutes.

A crack in the world

>>> Iceland is a spectacular country of extraordinary contrasts.

BELOW Strings of volcanic cones mark the 17-mile (27-km) long Laki eruptive fissure.

> This land of fire and ice squats upon the Mid-Atlantic Ridge, the plate boundary where fresh magma rises to make new oceanic crust that progressively and remorselessly pushes Europe and America apart. Like the tremendous scenery, Iceland's people are hard and rugged, seasoned to cope with most that nature can throw at them. Nothing, however, could have prepared them for the momentous events of June 1783.

Throughout the month, earthquakes had shaken the area of the Skaftá valley near Mount Laki to the south of the Vatnajökull ice cap in eastern Iceland, but the local population was used to this—Iceland was such a geological hotbed that it was never really still. As the first week of June came to an end, however, the shaking became far more intense and it was clear that something quite extraordinary was about to happen. At around 9:00 a.m. on June 8, the Earth literally tore itself apart. A spectacular 17-mile (27-km) long fissure system opened, spewing out incredibly fluid lava at a tremendous rate. Such was the pressure behind the molten rock that it was blasted into the air to form giant lava fountains as high as the Empire State Building. For three days, red hot lava continued to blast from the fissure, fragmenting and cooling to form ash clouds that were carried far and wide. Slowly the violence of the eruption receded, but only to be replaced by tremendous lava floods that poured far and wide across the surrounding countryside. The volume of the lava extrusion was unprecedented—a discharge rate comparable with the Amazon river. The lava soon tumbled into the 650 foot (200m) deep Skaftá valley, replacing a watery torrent with a seething river of molten rock. By the middle of July, the flow had poured over the precipitous

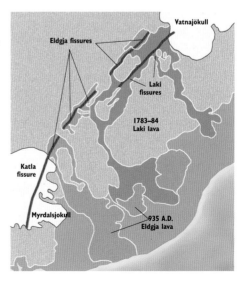

ABOVE Iceland is entirely built from lava that regularly spews from a chain of active volcanoes that slices right across the country.

LEFT Lavas pour from a fissure near Iceland's Krafla volcano.

BELOW Even bigger than the Laki event, the Eldgja fissure eruption split Iceland apart in 935 A.D.

Stapafoss Falls, forming a huge 12 mile (20km) wide lava delta on the flat coastal lowlands.

Long-term devastation

In just four days the lava flows had traveled 20 miles (35km), but the eruption continued until February of the following year. By the time the last flow ground to a halt, nearly 3 cubic miles (14,000cu.m.) of lava had been expelled, covering almost 200 square miles (600 sq.km)— roughly equal to the area of downtown Paris or London. Nearly 50 farms were entirely buried or badly damaged, and vast areas of agricultural land inundated. Despite their speed, the lava flows were still too slow to catch and overwhelm anyone and no deaths resulted from the direct effects of the eruption. But the longer-term and more insidious impacts were to prove a much greater threat to life and limb.

Lava and ash were not the only products of the Laki outburst. Huge quantities of noxious volcanic gases were also expelled, including sulfur dioxide, hydrogen sulfide, chlorine, and fluorine. A blue, foul-smelling haze covered the land, beneath which crops and pastures yellowed and died. Soon livestock began to sicken as they chewed on withered grass contaminated with fluorine. This deadly element took a terrible toll on the island's animals. Flesh fell from their bodies and manes and tails came off in the hand when pulled. Iceland's human population fared little better, with fluorosis causing their gums to swell, their hair to fall out, and boils and bony deformities to develop all over their bodies.

More than 75 percent of the island's sheep and horses and half of its cattle succumbed to poisoning or starvation—even the fish migrated temporarily from the polluted coastal waters. In the resulting famine, 10,000 people— roughly one-fifth of Iceland's population— starved. It took several years for food to become plentiful and many more before the "haze famine" became a distant memory.

Cataclysm!

>>> Southeast Asia teems with active and potentially active volcanoes, perhaps over 500 in all.

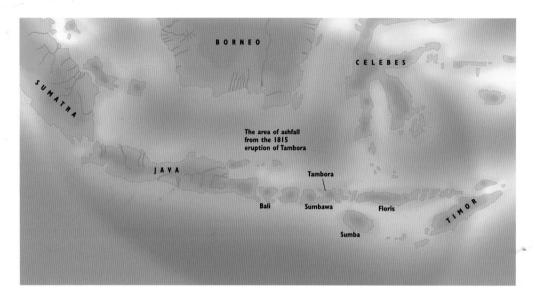

The area of ashfall from the 1815 eruption of Tambora

BORNEO · CELEBES · SUMATRA · JAVA · Bali · Sumbawa · Tambora · Floris · Sumba · TIMOR

LEFT The Great Tambora Eruption of 1815 dumped huge quantities of ash across much of Southeast Asia.
OPPOSITE A chain of cloud-shrouded active volcanoes stretches across the Indonesian island of Java and its neighbors.

> Most congregate in a narrow band coinciding with a destructive plate boundary that snakes its way through the Indonesian islands of Sumatra, Java, and Bali, before heading north—first into the Philippines and then Japan. In the nineteenth century, Indonesia hosted two of the most devastating and lethal eruptions of modern times. Most of us are familiar with the name of Krakatoa, the volcano that blew itself apart in 1883, but few have heard of the much greater eruption of Tambora nearly 70 years earlier—an event that was to take many more lives and have a much greater impact across the planet.

In 1811, Tambora was a small and innocuous volcano nestled peacefully on the island of Sumbawa, just to the east of Java and Bali. The following year, however, saw Tambora slowly awake, with earthquakes and steam explosions heralding the end of its long dormancy. For another three years the volcano grumbled and

smoked, on occasion causing concern among the local inhabitants, but no real fear. Many believed that because the activity had persisted for so long without a real eruption it was likely that things would soon die down again and the volcano return to its slumber. They could not have been more wrong. For the previous 36 months a gigantic volume of fresh magma had been gradually accumulating within and below Tambora, making ready to blast its way to the surface. The rumblings and shakings were a sign of this and would have been recognized as such by today's volcanologists, but in 1815 the signs meant nothing to the local people, who never anticipated the awful events that were to follow.

On April 5, the peace of the region was shattered by the first of a series of titanic explosions, as enormous pressures released the magma from its prison. Over the next month a series of gigantic explosions rocked the volcano, dumping ash and rocks on the surrounding countryside and at last terrifying the population. On May 10 and 11, the eruption reached its climax with a number of awesome detonations that blew the top of the volcano apart.

Catastrophic eruption

A graphic account of the cataclysmic phases of the eruption is provided by Sir Stamford Raffles, who, at the time, held the post of British Lieutenant Governor of Java. Raffles reported that the detonations were so extraordinarily loud that they could be heard in Sumatra nearly 1,000 miles (1,600km) away. He describes three spectacular "columns of flame" rising to immense heights and talks of the entire surface of Tambora being covered in incandescent material extending to great distances. Rocks, "some as large as the head," fell an amazing 6 miles (10km) from the volcano, while the torrential ash fall meant that total darkness reigned, even in Java, 300 miles (500km) to the west. The picture Raffles paints of the scene nearer the volcano is one of devastation, with buildings crushed and flattened by the weight of gray ash several feet deep, and the sea covered in so much floating pumice that ships could barely make any headway at all. Of the 12,000 local residents only 36 survived, the rest were either roasted alive in pyroclastic flows, battered by falling rocks, crushed in collapsed buildings, or washed away by tsunami.

And the terrible legacy of the eruption was not to end there. With the island buried beneath a thick mantle of ash, there was no prospect of a harvest that year, and in a few weeks starvation and disease began to exact a terrible toll on the survivors. With no external aid the situation deteriorated rapidly and within a year a further 82,000 men, women, and children had succumbed to the after-effects of the eruption, making Tambora the most lethal eruption ever recorded.

A HUNDRED WAYS TO DIE

Volcanoes can kill, maim, and destroy in a far greater variety of ways than other natural hazards such as floods and hurricanes—and all of these ways are extremely unpleasant.

Major volcanic blasts generate an impressive spectrum of destructive phenomena that can wipe entire cities and their inhabitants from the face of the Earth, burying any sign of their former existence beneath an immense blanket of ash, debris, and lava. For most people volcanoes are synonymous with flows of red hot lava, but in reality these are the least terrifying of all volcanic hazards and pose little threat to human life.

Less spectacular but far more disruptive are the clouds of volcanic ash that can blanket an area of thousands of square miles, bringing down power lines, collapsing buildings, and making driving almost impossible. Ash can also pose a serious threat to aircraft, contaminate water supplies, destroy crops, and cause health problems if inhaled.

Most terrifying and deadly of all volcanic hazards is the pyroclastic flow—a lethal cocktail of incandescent ash, superheated gases, and sometimes lava blocks the size of houses, that hurtles down the flanks of an exploding volcano with the force of a hurricane. Few people have survived the onslaught of a pyroclastic flow—these killers have claimed nearly 40,000 lives in the last 100 years alone. A close and equally deadly cousin is the lahar—a raging torrent of mud, generated by the mixing of water and volcanic debris. Only 15 years ago a huge tide of water and debris from the Nevado del Ruiz volcano buried the Colombian town of Armero,

enveloping its inhabitants in a muddy incarceration where 23,000 died.

And the awful effects of volcanic eruptions don't stop there. Gigantic landslides from coastal and island volcanoes produce huge and devastating sea waves that batter shorelines even thousands of miles away. Clouds of noxious gases suffocate local inhabitants and kill their livestock, while those made homeless by eruptions die through famine or by drinking contaminated water. On the grandest scale, a future super-eruption threatens to darken skies across the planet, plunging the world into the depths of a freezing volcanic winter from which our modern society may not emerge.

OPPOSITE Lavas trundle remorselessly toward evacuated homes on the flanks of Hawaii's Kilauea volcano.
BELOW Large, hot rocks hurled from the Caribbean island of Montserrat's Soufrière Hills volcano in 1996 take their toll on houses in the nearby settlement of Long Ground.

Rivers of lava

>>> Volcanoes can be classified in many different ways, but perhaps the most straightforward is to divide them into "red" volcanoes and "gray" volcanoes.

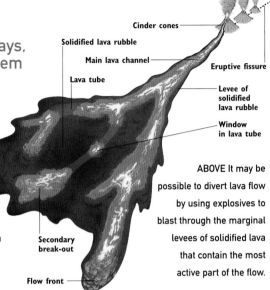

Cinder cones
Solidified lava rubble
Main lava channel
Lava tube
Eruptive fissure
Levee of solidified lava rubble
Window in lava tube
Secondary break-out
Flow front

ABOVE It may be possible to divert lava flow by using explosives to blast through the marginal levees of solidified lava that contain the most active part of the flow.

> While the latter erupt explosively, producing clouds of gray ash, activity at the former is dominated by glowing red lava flows, which ooze relatively quietly out of the ground and creep downslope at walking pace or less. Such lava-dominated volcanoes are said to be effusive rather than explosive and include Mount Etna in Sicily, Kilauea and Mauna Loa on the island of Hawaii, and Krafla and Hekla in Iceland.

Temperatures approaching 1800°F (1000°C) or more—four times hotter than the maximum heat of a kitchen oven—ensure that a lava flow will incinerate anything in its path, while its enormous strength will reduce to rubble any buildings that get in its way. Lavas are generally much too sluggish, however, to pose any serious threat to life, and there is always plenty of time for evacuation. Sometimes there is sufficient respite to move possessions too, and during the 1928 eruption of Etna the inhabitants of Mascali moved everything but the buildings themselves before the town was bulldozed and buried by the advancing flows. A little over 250 years earlier, the inhabitants of nearby Catania fared worse when lavas from one of the most devastating of Etna's historical eruptions burst through the city walls and laid waste much of the city.

The eruption also provided the opportunity for one of the first lava diversion attempts. As the flows lapped against the city walls, the city's bravest souls attacked the margin of the lava flow with picks and axes. After a long battle they succeeded in breaching a hole in the solidified flow margin, allowing the molten interior to

burst through. As lava poured out and away from the city, however, disgruntled inhabitants from nearby Paternò appeared, concerned that the diverted flow would hit their town. The Catanese were driven off, the breach healed itself, and the flow resumed its original course. Soon the protecting walls were leveled and over 80 percent of the city was destroyed, leaving 20,000 homeless. During recent eruptions of Etna, scientists have again tried to divert flows

BELOW Lava from Mauna Ulu on Hawaii's Kilauea volcano enters the sea on the west of Apua Point. BOTTOM RIGHT Lavas from the 1983 eruption of Mount Etna in Sicily engulf the home of an unfortunate resident.

away from inhabited areas, and in 1992 a successful diversion may have helped to limit damage to the threatened town of Zafferana.

Although volcanoes like Etna continue to pose a constant threat to local communities, this does not dissuade people from living in close proximity to them. The fertile volcanic soils support lucrative agriculture and for this reason over 20 percent of the population of Sicily lives on Etna's slopes. On average, every point on the volcano is covered by lava every 400 years, so 15 generations of a family can exist peacefully on its flanks before seeing their farm and land inundated by lava—clearly a risk worth taking.

Colossal outpourings

While awe-inspiring in their own right, the lavas of Mount Etna are mere dribbles when compared with the molten extrusions of other effusive volcanoes. In 1783, a 17 mile (27km) fissure at Lakagigar (Laki) in Iceland spewed out thousands of times more lava than even the greatest of Etna's efforts; 50 years earlier, in 1730, the peace of Lanzarote (Canary Islands) was shattered by a gigantic expulsion of lava and ash that lasted for six years. By the time the last flows had cooled and solidified, over one-quarter of the island—around 50 square miles (200sq.km)—had been buried, destroying towns and villages and turning once fertile farmland into a sterile volcanic desert. The biggest lava eruptions—known as flood basalts—dwarf even these. In Siberia, the northwest United States, India, and elsewhere in the world, colossal outpourings of lava have covered regions larger than the United Kingdom. The Deccan Traps lavas of northwest India cover 120,000 square miles (500,000sq.km) and have thicknesses of up to 1.2 miles (2km). This unimaginable expulsion of lava occurred 65 million years ago, and is held responsible by some for the demise of the dinosaurs.

RIGHT Fountains, cascades, and bubbles of lava pour from the east flank of Mauna Ulu on Hawaii's Kilauea volcano, ultimately entering the sea in a spectacular explosion of steam.

Hurricanes of fire

>>>Just before 8 a.m. on Ascension Day, 1902, the bustling French colonial town of St. Pierre on the Caribbean island of Martinique was obliterated: wiped from the face of the Earth by that most feared of all volcanic phenomena—the pyroclastic flow.

>Mont Pelée had been rumbling more and more violently throughout April and by early May heavy ash fall and violent shaking of the ground increased concern among the islanders, many of whom fled in panic to the false security of St. Pierre. It was to be the greatest and last mistake of their previously tranquil lives. The local newspaper, *Les Colonies*, in an attempt to calm a growing feeling of desperation among the town's inhabitants and refugees, announced:

"Mont Pelée is no more to be feared by St. Pierre than Vesuvius is feared by Naples. Where could one be better off than St. Pierre?"

Of course the population of Naples has always feared Vesuvius, and the inhabitants of St. Pierre were right to fear the wrath of the volcano that towered over the town.

The sunny, Caribbean morning of May 8 was greeted by a colossal blast from the volcano directed straight at the town, as if from the barrel of a shotgun. Within a minute a terrible glowing hurricane of molten fragments and superheated steam crashed into St. Pierre and enveloped it in blistering darkness. When the volcanic cloud finally cleared, it was to reveal a scene of utter devastation. Virtually no buildings were left standing, heavy cannons had been torn from their moorings, and a statue of the Virgin Mary, weighing over three tons (756kg), had been hurled 65 feet (20m) from its base. Even

LEFT Volcanologists take the temperature of new pyroclastic flow deposits on the Soufrière Hills volcano. Note the lava dome in the background.

ABOVE Images of St. Pierre on the Caribbean island of Martinique, before and after the devastating eruption of Mont Pelée in 1902.

on the sea there was no respite, and in the harbor, 15 ships drifted, crippled and burning like their crews. An eerie silence reigned, for hardly anyone remained alive. Virtually all 29,000 people in the town on that fateful morning were immolated within seconds—the clothes ripped from their bodies, the flesh burned from their bones, and their organs ruptured by the explosive expansion of the water they contained. Miraculously Auguste Ciparis, a prisoner incarcerated in a thick-walled stone cell, survived his terrible burns and went on to relive his ordeal for the pleasure and entertainment of visitors to Barnum & Bailey's Circus. Ciparis and one other were the only survivors.

Glowing killers

Pyroclastic flows, also known by the French term *nuées ardentes* (literally "glowing avalanches"), are one of the biggest killers generated by volcanoes. They were responsible for wiping out many of the inhabitants of Pompeii nearly 2,000 years ago, and only 24 hours before the St. Pierre catastrophe they had taken another 1,700 lives on the neighboring Caribbean island of St. Vincent. The destructive capability of a pyroclastic flow derives from a combination of high temperature, sometimes up to 1400°F (800°C), high velocity, typically in excess of 60 miles (100km) per hour, and the masses of volcanic debris often contained within. Few structures can withstand the impact of a pyroclastic flow, and images of St. Pierre after the eruption are barely distinguishable from those taken of Hiroshima and Nagasaki following their destruction by atomic bombs. A pyroclastic flow's high velocity makes it impossible to outrun, the superheated gases providing a

near-frictionless cushion on which the rest of the flow can ride. Pyroclastic flows can be formed in a number of ways, bursting sideways from beneath a plug of old, solidified lava—as at Mont Pelée—or from the collapse of a lava dome or eruption column.

Only a few years ago, in 1997, 19 inhabitants of Montserrat were killed by pyroclastic flows caused by the collapse of the Soufrière Hills lava dome. Six years earlier, only the evacuation of 250,000 people prevented pyroclastic flows formed from the collapse of the Pinatubo eruption column from causing the wholesale slaughter of the local population.

ABOVE Although horribly burnt, prisoner Auguste Ciparis was saved from death during the eruption by the thick stone walls of his prison cell.

BACKGROUND Pyroclastic flow, made up of clouds of ash and steam, is the ultimate volcanic hazard, killing or obliterating everything in its path.

Torrents of mud

>>> Armero was a town waiting to die. Although nestled in Colombia's Lagunillas river valley, a seemingly comfortable 30 miles (50km) from the towering, ice-covered Nevado del Ruiz volcano, it was far from safe from the volcano's anger.

> The valley offered a perfect conduit for the transport of volcanic debris, which, during eruptions, poured down from Ruiz's snowy summit as a torrent of mud known as a lahar. Time and again these slimy killers had wiped out settlements in the valley floor. In November 1984, Ruiz rumbled back into life after nearly 400 years of quiet, and it continued to shake and spew small amounts of ash until September of the following year. While the local population and the civil authorities were lulled into a false sense of security by the minor nature of the activity, geologists were becoming increasingly concerned. This turned to alarm in mid-September 1985, when a small eruption melted part of the summit ice field generating a 12-mile (20-km) long mudslide. Still no action was taken, and the scene was set for the second greatest

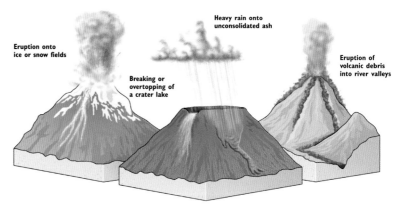

ABOVE Lahars can be generated in many different ways and may form even when a volcano is not in eruption.
BELOW LEFT The Pinatubo blast dumped huge quantities of ash and pyroclastic flow deposits on the volcano's flanks—a ready source for lahars.
RIGHT Memorial to those who lost their lives in the Christmas Eve 1953 rail disaster caused by lahars from New Zealand's Mount Ruapehu volcano.

volcanic disaster of the twentieth century.

Just after 9:00 p.m., Ruiz exploded into life, blasting pyroclastic flows across the summit ice cap causing catastrophic melting. Water poured into the upper reaches of the Lagunillas river at more than 1 million cubic feet (30,000cu.m) per second—enough to drown central London or downtown Los Angeles in water 300 feet (100m) deep. Scouring the valley sides, the flood soon collected sufficient debris to turn it to a torrent of watery mud, hurtling downhill at 30 miles (50km) an hour. Two hours later, a wave of mud 130 feet (40m) high crashed into Armero and neighboring settlements. There was no escape, and as the flow subsided the mud

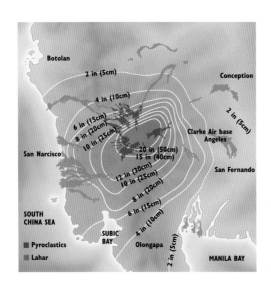

THIS MONUMENT MARKS THE SITE OF THE TANGIWAI RAIL DISASTER ON THE NIGHT OF 24TH DECEMBER 1953 WITH THE TRAGIC LOSS OF 151 LIVES

ERECTED IN JUNE 1989
BY THE LIONS AND LIONESS
CLUBS OF RUAPEHU, WAIOURU,
TAIHAPE AND HUNTERVILLE
WITH ASSISTANCE FROM
NZ TIMBERLANDS LTD.
NZ MASTER MONUMENTAL MASONS ASSOC. INC.
NATIONAL RADIO
NZ RAILWAYS CORPORATION
DEDICATED ON THE 18TH JUNE 1989

quickly dried to form a cement-like tomb from which over 23,000 people never emerged alive.

Lahars are not only formed by eruptions onto or through snow and ice, but can also result from the breaching or overtopping of a crater lake or simply from heavy rain on unconsolidated ash. At Kelut volcano (Indonesia), violent eruptions sometimes eject water from the summit crater lake, generating lahars that pour down the sides of the cone, and in 1919 such a flood killed over 5,000 people and obliterated more than 100 villages.

Deadly cascade

Devastating lahars can even be generated without any eruptive activity at all. On Christmas Eve in 1953, the wall of the Mount Ruapehu (New Zealand) crater lake collapsed sending a deadly cascade of water, ice, mud, and volcanic debris tumbling down the Whangaehu river valley. Within minutes the boiling surge had pounded into a railway bridge downstream, turning it to kindling and twisted metal. Just two hours later tragedy struck as the Wellington–Auckland night express hurtled through space and into the raging torrent below. Fewer than half of the 285 passengers survived. More recently, in 1998, the torrential rain associated with Hurricane Mitch triggered the collapse of part of the Casita volcano in Nicaragua, forming a 12-mile (20-km) long lahar that buried several towns and killed 2,000.

The lahar threat may not necessarily subside once an eruption ceases. Following the huge 1991 eruption of Pinatubo in the Philippines, remobilization of pyroclastic flow and ash deposits by torrential tropical rains continued to generate lahars that clogged rivers, inundated farmland, and flooded cities. For the last decade the authorities have fought a never-ending battle against the mud by continuously dredging and deepening the rivers that drain the volcano and

constructing levees along the banks to contain the muddy torrents.

Where will lahars strike next? Perhaps Mount Rainier in Washington State. This huge ice-covered giant looming over the cities of Seattle and Tacoma is poised, during its next eruption, to send rivers of mud pouring into the towns surrounding the volcano. Emergency plans are in place to cope with the threat, but the local population remains forever watchful and wary.

ABOVE In Colombia, the town of Armero and 23,000 of its inhabitants are buried beneath a thick blanket of mud from Nevado del Ruiz.

Deadly clouds

>>> In stark contrast to the red volcanoes that ooze lava, the gray ash-forming volcanoes are considerably more violent and, therefore, far more dangerous.

> Gray volcanoes blast magma into the atmosphere in the form of explosions that physically tear the hot, plastic magma into fragments known as pyroclasts. Depending upon the power of the explosion, these can range in size from blocks as big as houses through head-sized, torpedo-shaped volcanic "bombs" to fine, talcum-powder like ash. Clearly, anything that goes up from a volcano must eventually come down, and when it does—whatever its size—it is known by the collective term "tephra."

Suffocating layer of ash

The distance a volcano can hurl objects depends upon their size. The larger blocks follow ballistic trajectories—like balls from a cannon—and rarely travel more than a few miles. These only pose a threat to people and structures living close to an active vent. Finer ash particles, on the other hand, can be carried by the wind for thousands of miles before slowly drifting down to the surface. As a result the ash component of tephra affects a much greater area and is far more disruptive. The volumes of material ejected from an exploding volcano can be difficult to imagine. For example, the great Tambora eruption of 1815 blasted out sufficient ash to blanket an area of 965,000 square miles (2.5 million sq.km).

More recently, in 1980, ash from the explosion generated by the collapse of Mount St. Helens covered much of the northwest United States, from Seattle in the west to Montana in the east. Just over a decade later, the 1991 Pinatubo eruption spewed over 1.7 cubic miles (7 cu.km) of ash and debris—enough to bury central London under a thick, gray, blanket two-thirds of a mile (1km) thick. Falling as it did during the coincidental passage of Typhoon Yunya, much of the Pinatubo ash combined with torrential rain as it fell, forming thick, sticky mud balls. The weight of this sodden, muddy ash, as it accumulated on buildings and other structures, caused their roofs to collapse. At the nearby Cubi Point Naval Air Station, the ash rapidly accumulated on exposed aircraft, causing them to tilt under the weight. Aircraft in the air are also under particular threat from volcanic ash, and

BELOW Heavy volcanic ash fall from the 1991 eruption of Pinatubo in the Philippines severely damaged aircraft at Cubi Point Naval Air Station.

there have been over 80 encounters between aircraft and ash clouds in the last 20 years. Ash not only "crazes" the windscreen, leading to dramatically reduced visibility, but clogs the engines causing them to stop and the plane to lose power. None have crashed yet but there have been some close calls—most notably in 1982, when a British Airways 747 fell 26,000 feet (8,000m) as the pilot struggled to regain the use of its engines following an encounter over Indonesia with a huge column of ash from the Galunggung volcano.

As well as causing damage to buildings, falling ash seriously disrupts communications. Power and telephone lines are soon brought down during heavy ash fall, while driving is nearly impossible due to pitch darkness. The rate of ash accumulation during the biggest eruptions is hard to believe. At the height of the 1883 Krakatoa blast, for example, the deluge of hot ash accumulated at rates of up to 6 inches (15cm) in ten minutes—almost 3 feet (1m) an hour!

Volcanic ash also leads to longer-term problems. A thick blanket of ash will destroy crops, preventing plants from photosynthesizing. If it carries with it noxious volcanic compounds, such as fluorine, ash can also poison fisheries and contaminate drinking water and animal feed, leading to the wholesale loss of livestock that afflicted Iceland after the 1783 eruption of Laki. The long-term presence of ash, as at Montserrat, in the Caribbean, or around the Rabaul volcanoes in Papua New Guinea, can have a particularly damaging effect on human health, as the fine dust exacerbates or causes respiratory illnesses such as asthma and silicosis.

ABOVE This house was perched on the flanks of Mount Pinatubo. Only rarely will a building survive beneath a blanket of ash this thick.

Collapsing volcanoes

>>> The snow-capped volcanoes of the Cascade Range in the northwestern United States tower over the surrounding plains like invincible sentinels, exuding impressions of solidity and permanence, suggesting they will remain unchanged for ever.

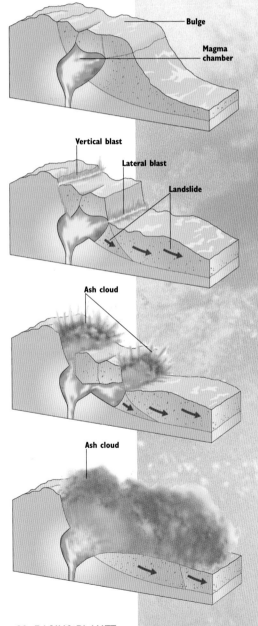

Bulge

Magma chamber

Vertical blast

Lateral blast

Landslide

Ash cloud

Ash cloud

LEFT (TOP TO BOTTOM) The climactic eruption of Mount St. Helens in 1980, was triggered by a series of gigantic landslides that removed the northern flank in less than a minute. **BACKGROUND** The collapse of Mount St. Helens triggered a huge explosive eruption that lasted for many hours and killed 57 people.

> This could not be further from the truth. Many volcanoes are in fact rotten to the core; weak piles of ash, lava, and rubble that—given half a chance—will collapse into a deadly, churning mass of volcanic debris.

In March 1980, Mount St. Helens in Washington State, U.S.A.—one of the most spectacular of the Cascade volcanoes—coughed back into life following over 120 years of sleep. Earth tremors, fresh cracks in the summit region, and explosions of ash and steam from new vents heralded the start of a new eruptive cycle that was to last for a decade. As the first volcanic eruption in the 48 contiguous states during the era of television, the unfolding events at Mount St. Helens captured the rapt attention of the world's media, which descended on the area.

Throughout March and April, as cameras whirred and journalists scribbled, fresh magma continued to feed the volcano. Because of a thick plug of solidified lava emplaced during the previous eruption, however, the new magma was unable to burst through the summit. Instead, it was forced to push beneath the north flank of the volcano, progressively bulging it outward like a giant, festering boil. By early May the "boil" was 1.2 miles (2km) across and 330 feet high, and growing at over 3 feet (1m) a day. At 8:32 a.m. on May 18, it finally burst. By now the flank was so unstable that a small earthquake was sufficient to detach the bulge from the rest of the volcano, causing it to slide off as a gigantic rock avalanche. As the weight of the north flank was removed, pressure on the pent-up magma beneath was released causing a cataclysmic blast powerful enough to flatten over 200 square miles (600 sq.km) of forest and be heard more than 180 miles (300km) from the volcano. Devastating pyroclastic flows quickly overtook the rock avalanche, while, almost immediately, the collapsing debris mixed with available lake and river water, forming boiling lahars that demolished roads, bridges, and homes. Ash continued to blast out of the volcano for several hours, falling as far as Montana, 600 miles (1,000km) away. When things finally calmed down, 57 people were dead, including volcanologist David Johnston, and over $1 billion of damage had been incurred. The impression of volcanoes as solid, stable, objects was lost for ever as watching scientists, film crews, and the world at large were staggered by the destructive power of a collapsing volcano.

Vicious life cycle

Mount St. Helens triggered considerable scientific interest in how and why volcanoes collapsed and sent volcanologists trawling through records of past eruptions. It soon became apparent that the collapse of all or a fraction of a volcanic edifice was a normal part of a volcano's life cycle, which happened somewhere on the planet perhaps four times a century. Examples from the past came to light (see table above). In 1640, part of the Japanese volcano Komaga-Take collapsed, killing 700 people and destroying many ships. Between 1700 and 1900, three deadly landslides occurred at Japanese volcanoes, together accounting for almost 17,000 lives. Further west in Java, in 1772, a huge landslide removed the northeast flank of the Papandayan volcano, and within minutes a torrent of rock and debris destroyed 40 villages, killing nearly 3,000 of their inhabitants. Just over a century later the Ritter Island volcano off the coast of New Britain (Papua New Guinea) also fell apart, sending lethal tsunami crashing onto settlements along

Volcano collapse events

VOLCANO	YEAR	DEATHS	NOTES
KOMAGA-TAKE (JAPAN)	1640	700	DEATHS DUE TO TSUNAMI
OSHIMA-OSHIMA (JAPAN)	1741	1,475	EARTHQUAKE TRIGGERED COLLAPSE; DEATHS DUE TO TSUNAMI
PAPANDAYAN (INDONESIA)	1772	3,000*	40 VILLAGES OBLITERATED
UNZEN (JAPAN)	1792	14,528	DEATHS DUE TO TSUNAMI
BANDAI-SAN (JAPAN)	1888	500*	SEVERAL VILLAGES BURIED
RITTER ISLAND (PAPUA NEW GUINEA)	1888	3,000*	DEATHS DUE TO TSUNAMI
BEZYMIANNY (RUSSIA)	1956	NONE	CARBON-COPY OF MOUNT ST. HELENS
MOUNT ST. HELENS	1980	57	FIRST OBSERVED VOLCANO COLLAPSE

*APPROXIMATE TOTAL

adjacent coastlines. In 1956, barely a quarter of a century before the Mount St. Helens blast, Bezymianny volcano on the Russian Kamchatka peninsula blew itself apart in an almost carbon-copy eruption. Unlike Mount St. Helens, the event was far from the media spotlight, and few outside the Russian volcanological community were aware of the dramatic fireworks among the snowy wastes of eastern Russia.

ABOVE As the north flank slid away, pressure on the underlying magma was instantaneously released, generating a blast wave that flattened over 200 square miles of mature forest.

Making waves

>>>When the gods of molten magma and the sea—Vulcan and Neptune—conspire together, the outcome can be truly appalling.

Volcanic tsunami

VOLCANO	YEAR	CAUSE	DEATH TOLL	NOTES
KOMAGA-TAKE (JAPAN)	1640	LANDSLIDE	700	
SANTORINI (GREECE)	1650	ERUPTION	50	
LONG ISLAND (PAPUA NEW GUINEA)	1660	ERUPTION	APPROX. 2,000	TSUNAMI AND PYROCLASTIC FLOWS
GAMKANORA (INDONESIA)	1673	ERUPTION	MANY	
OSHIMA-OSHIMA (JAPAN)	1741	LANDSLIDE	1,475	
UNZEN (JAPAN)	1792	LANDSLIDE	14,528	
TAMBORA (INDONESIA)	1815	ERUPTION	MANY	10,000 KILLED BY DIRECT EFFECTS OF ERUPTION
RUANG (INDONESIA)	1871	LANDSLIDE	400	COLLAPSE OF LAVA DOME
KRAKATOA (INDONESIA)	1883	ERUPTION	36,417	MOST KILLED BY TSUNAMI
RITTER ISLAND (PAPUA NEW GUINEA)	1888	LANDSLIDE	APPROX. 3,000	WAVES 40–50 FEET (12–15M) HIGH
TAAL (PHILIPPINES)	1965	ERUPTION	>200	MOST DROWNED DUE TO BOATS CAPSIZING
ILIWERUNG (INDONESIA)	1979	LANDSLIDE	539	WAVES 30 FEET (9M) HIGH

>Just after 10:00 a.m. on August 27, 1883, their scheming resulted in one of the greatest of all volcanic catastrophes. Having slumbered for many years, the small volcanic island of Krakatoa, guarding the Sunda Strait between Java and Sumatra (Indonesia), at last jolted back into life at the end of May. Loud explosions rocked the region throughout June and July causing increasing anxiety among the inhabitants of Batavia (now Jakarta) and neighboring towns, but things did not really liven up until midday on August 26. Then, a series of titanic explosions was heard all over Java as the volcano blasted a cloud of black ash to a height of 15 miles (25km). All night the eruption continued, trapping thousands of terrified people in homes that were becoming buried under a rain of hot pumice. But much worse was to come. At dawn, the already shell-shocked inhabitants were greeted by the loudest explosion ever heard as Krakatoa blew itself apart with the force of a million Hiroshima bombs. The report was heard as far away as Alice Springs in Australia and Rodriguez Island in the Indian Ocean, both 1,800 miles (3,000km) from Krakatoa.

The cataclysmic blast resulted from the evacuated magma chamber beneath the volcano collapsing, allowing trillions of gallons of seawater to mix with the molten magma. As the sea poured into the gigantic submarine void, it vanished from the shores of Java and Sumatra, leaving dying fish flapping on empty sands and boats stranded in the harbor. Minutes later, however, it was back with a vengeance, as towering waves up to 130 feet (40m) high scoured over 160 villages and over 36,000 men, women, and children from the islands. The terrible mayhem was described by engineer N. van Sandick of the Dutch ship *Loudon*:

"Like a high mountain, the monstrous wave precipitated its journey to the land. Immediately afterwards another three waves of colossal size appeared. And before our eyes this terrifying upheaval of the sea, in a sweeping transit, consumed in one instant the ruin of the town. There, where a few minutes ago lived the town of Telok Belong, was nothing but the open sea."

Killer waves

Such huge sea waves, or tsunami, often result from submarine earthquakes, but they are also the source of some of the highest death tolls involving volcanic activity, taking at least 55,000 lives over the last 350 years. Problems can arise wherever volcanoes and large bodies of water —either the sea or a lake—coexist. Volcanic tsunami are strongly linked to volcano collapse, and on six occasions gigantic volcanic landslides

into the sea or a lake have been responsible for lethal waves. On average, such tsunami-generating collapses occur in Japan every century or so, with the collapse of Unzen in 1792 resulting in the second most lethal volcanic tsunami recorded. Most recently, a huge landslide from the Iliwerung volcano on the island of Lomblen (eastern Indonesia) plummeted into the sea, generating tsunami 30 feet (9m) high that obliterated four villages and killed over 500 inhabitants.

Even those living far from the sea can be affected. Taal volcano in the Philippines is an island in the middle of a large lake, and also highly active, with two lethal eruptions in the past 100 years. The second of these, in 1965, sent the panicked inhabitants of the island scrambling for their boats and the safety of the mainland. They never made it. The shock waves generated by the violent volcanic explosions spawned tsunami that sped outward to catch and capsize the boats of the fleeing islanders, drowning over 200.

ABOVE Tsunami following the 1883 eruption of Krakatoa killed thousands around the shores of Java and Sumatra.
BELOW Such is the power of tsunami that boats can be carried far inland to be left high and dry when the sea withdraws.

VULCAN'S WEATHER

Volcanic eruptions differ from other natural disasters in that even quite small volcanic events have the ability to affect the entire planet.

This extraordinary reach arises from the fact that the gas and dust ejected during a volcanic blast is capable of modifying the Earth's climate. In 1982, a modest explosive blast at the El Chichón volcano in Mexico generated pyroclastic flows that killed nearly 2,000 people, but the effects of the eruption did not stop there. Clouds of sulfur-rich gases entered the stratosphere and rapidly spread across the planet, forming a veil that blocked out some of the Sun's radiation and reduced surface temperatures. On this occasion the temperature fall was small—around 0.5°C—and its effect was minimal, but larger volcanic eruptions in the past have had a much more devastating impact on the world's climate.

During the past 250 years, eruptions in both Iceland and Indonesia have seriously affected the climate of the northern hemisphere, causing a noticeable cooling and damaging both crop yields and human health. Looking back further, it appears that the largest eruptions of all are capable of lowering temperatures to such a degree that the planet is, temporarily at least, plunged into a deep freeze from which it does not emerge for years.

Furthermore, we do not yet fully understand how volcanic eruptions can affect the climate. Although their cooling effect has been recorded on several occasions, this is not always the outcome. The weather machine is so complex that it is difficult to predict how it will respond

to a large input of volcanic material. Recent studies have in fact revealed that, following large explosive eruptions, some parts of the world cool down while others warm up. There has even been some speculation that the meteorological phenomenon known as El Niño, which involves a pronounced warming of the eastern Pacific Ocean, and has an extreme effect on the weather across the planet, may be triggered by volcanic eruptions, although others dispute the link. Clearly, volcanoes and climate are bound intimately together, and in a world where meteorological hazards such as floods and hurricanes are becoming increasingly frequent and destructive, we ignore the volcanic dimension at our peril.

OPPOSITE Volcanic ash, dust, and gas from large explosive eruptions can shroud the entire Earth, cutting out solar radiation and reducing global temperatures.
TOP In just ten weeks, sulfur dioxide gases from the 1991 Pinatubo eruption in the Philippines spread right across the planet.
ABOVE Ash falls like snow during the 1991 eruption of Mount Pinatubo.

A chilling veil

>>> The first person to give serious thought to the climatic impact of large volcanic eruptions on the planet was H. H. Lamb who, in 1963, while working at the U.K. Meteorological Office, devised the Dust Veil Index.

> The Dust Veil Index (DVI) records the amount of fine ash pumped into the stratosphere by eruptions and transported around the Earth, by measuring how opaque the atmosphere is to solar radiation—the higher the opacity the larger the DVI. Lamb determined that there was a correlation between such high DVI values and global cooling, and he supposed that the cooling effect resulted from fine volcanic dust cutting out a certain proportion of the Sun's radiation. We now know, however, that even fine dust does not remain in the stratosphere (which starts 6 miles [10km] above the surface) long enough to have a significant cooling effect.

The real culprit is not dust but the sulfur gases, hydrogen sulfide and—especially—sulfur dioxide, that are pumped into the atmosphere in huge quantities during large volcanic eruptions. Once they reach the stratosphere, such gases combine with available water vapor to form a fine mist of sulfuric acid aerosols (microscopic droplets) that is particularly effective at reflecting

LEFT Solar radiation entering a cloud of volcanic aerosols is reflected, scattered, and absorbed, causing cooling of the troposphere.

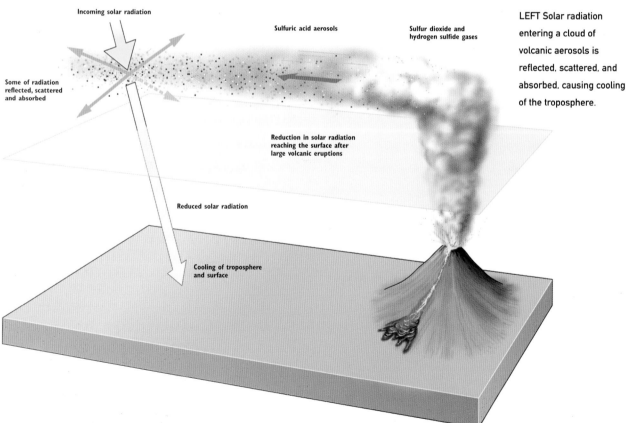

Incoming solar radiation

Sulfuric acid aerosols

Sulfur dioxide and hydrogen sulfide gases

Some of radiation reflected, scattered and absorbed

Reduction in solar radiation reaching the surface after large volcanic eruptions

Reduced solar radiation

Cooling of troposphere and surface

ABOVE Volcanic dust and gas in the atmosphere can lead to brilliant sunrises and sunsets.

solar radiation back into space. The aerosols are spread over the Earth by stratospheric winds forming a haze that increases the reflectiveness, or albedo, of the planet.

Optical illusions

The aerosol cloud can also have other effects and large eruptions such as those at the Indonesian volcanoes Krakatoa in 1883 and Tambora in 1815 were both followed, all over the world, by brilliantly colored sunrises and sunsets and by strange optical effects such as blue Moons and brown Suns. Surprisingly, it is not just the size of a volcanic eruption that determines the scale of its impact on the climate—the composition of the magma is also important: eruptions of sulfur-rich magma have the potential for greater global cooling. Furthermore, the location of the volcano is critical: eruptions in the tropics have a better chance of sending aerosol clouds into the wind systems of both northern and southern hemispheres. Eruptions at higher latitudes are far more likely to affect the climate of the hemisphere they occupy. The impact of such temperate or polar eruption may, however, be

BELOW LEFT Significant falls in solar radiation reaching the Earth's surface were observed after the eruptions of Agung, El Chichón, and Pinatubo.

ABOVE At ground level, noxious volcanic gases can damage crops and livestock, and lead to respiratory problems— even death.

more dramatic as the stratosphere becomes lower with increasing latitude.

Volcanic gases also cause problems at lower levels in the atmosphere, although these are largely confined to the immediate vicinities of volcanoes. At Poás in Costa Rica, hydrogen sulfide and sulfur dioxide, expelled from open vents in the crater, have been implicated in crop damage around the volcano and an increase in respiratory illness.

Carbon dioxide gas is also a problem. Twice in the mid-1980s, the sudden overturn of the waters at two crater lakes in the Cameroon resulted in carbon dioxide gas pouring across the surrounding countryside, asphyxiating thousands of local inhabitants and their livestock.

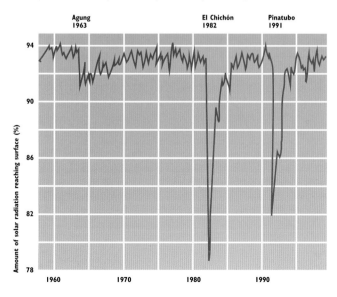

Agung 1963   El Chichón 1982   Pinatubo 1991

Amount of solar radiation reaching surface (%)

94

90

86

82

78

1960   1970   1980   1990

The noxious cloud

>>>The worst natural disaster ever to hit Iceland was without a doubt the Laki fissure eruption of 1783, decimating the livestock population and leading to widespread famine that killed one in five of the island's inhabitants.

>The impact of the eruption was not, however, to stop there. As the great fissure system continued to spew forth lava, a blue haze of noxious sulfuric acid aerosols gradually spread across Iceland and from there to mainland Europe, North America, North Africa, and western Asia. Just two weeks after the start of the eruption huge quantities of the aerosols became trapped in the column of rising air of a low pressure system over Iceland, from which high altitude winds carried them southward toward Britain and Europe. At the time the region was dominated by a high-pressure system; an area of more dense, descending air that returned the aerosols back to low altitudes. Under the calm surface conditions, they spread out to form a smoky fog that hung over the landscape throughout the summer and well into the winter. First-hand descriptions of the aerosol fog's effect in Europe were provided from Paris by Benjamin Franklin who, at the time, was the first diplomatic representative in France of the fledgling United States. He reported:

"During several of the summer months of the year 1783 there existed a constant fog over all Europe and a great part of North America. This fog was of a permanent nature; it was dry, and

FAR LEFT The prevailing weather conditions carried the noxious cloud to Europe and held it there for many months.
BELOW LEFT Winter temperature records from the eastern U.S. reveal a sharp drop in temperatures after the Laki eruption.

BACKGROUND Following the 1783 Laki eruption in Iceland, a bitter winter gripped both Europe and North America.

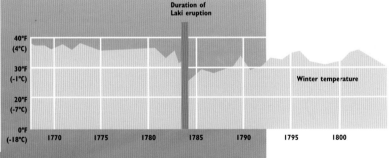

the rays of the sun seemed to have little effect towards dissipating it. They were indeed rendered so faint in passing through it that when collected in the focus of a burning glass, they would scarce kindle brown paper."

The renowned British naturalist, Gilbert White, also observed the fog, and described the Summer of 1783 as:

"…an amazing and portentous one, and full of horrible phenomena…the peculiar haze or smoky fog, that prevailed for many weeks in this island and in every part of Europe…was a most extraordinary appearance."

Sulfurous fog

The concentration of volcanic gases at low levels in the atmosphere had a devastating impact on agriculture. In Scotland, barley, oats, and rye became brown and withered, while conifers such as Scot's fir and larch were also damaged. Further south, the wheat crop suffered similar damage and across Europe the leaves of many plants became dried and blackened. Insects fared badly, and aphids and other species were exterminated in huge numbers. People fared little better—the strongly sulfur-smelling fog is reported to have caused headaches, asthma, and general breathing difficulties. Recent research of parish records in Britain has revealed a higher than normal death rate, which may reflect the impact of the foul-smelling fog on the young, old, and infirm. Reports of a thick, hot fog also came from Spain, southern France, and Italy, where the haze was so thick that barge pilots found it impossible to navigate without a compass.

Even with summer's end there was to be no relief as the stagnating sulfurous fog blocked out the Sun's rays and led to the early onset of winter. Temperatures in the northern hemisphere are estimated to have fallen by around 2°F (1°C), leading to a bitterly cold winter in Europe and North America. Average winter

temperatures in the eastern United States dropped almost 8°F (5°C) lower than normal, while in Europe sharp early frosts froze the surface and successive snowfalls remained unmelted for months.

The impact of around 100 million tons of volcanically produced sulfuric acid aerosols on the northern hemisphere was devastating but far from unique. Accounts of similar aerosol fogs can be found in literature dating back over 2,000 years. Plutarch, writing in 44 B.C., talks of a "pale sun without radiance" that caused fruits to become "withered away and shriveled up on account of the coldness of the atmosphere." Most intriguing of all is the so-called "mystery fog" of 536 A.D., which is reported to have led to cattle dying and humans suffering illness—but for this a volcanic source has yet to be identified.

ABOVE In addition to poor weather and serious crop damage, the volcanic smog also seems to have led to premature death among the old, infirm, and very young.

The year without a summer

>>> In the history books, the year 1815 is best known for the game of cat and mouse played across Europe by the armies of Napoleon and Wellington and culminating in the battle of Waterloo.

> Another event that year, however, was also to have a major impact on the continent. Nearly 7,500 miles (12,000km) away in far off Indonesia, the Tambora volcano blew itself apart in one of the greatest eruptions of recent times. The immediate and short-term effects were the destruction of the Indonesian island of Sumbawa and the deaths of 92,000 of the island's inhabitants. Another 12 months were to elapse, however, before Europe—and indeed North America—were to feel the somewhat delayed consequences of Vulcan's fury.

As well as ejecting over 24 cubic miles (100cu.km) of ash and debris, the Tambora blast also pumped around 200 million tons of sulfur gases into the atmosphere, where they combined eagerly with water vapor to form 150 million tons of tiny sulfuric acid droplets. As the year ebbed, these aerosol clouds spread northward in the stratospheric winds, forming a veil across the Sun and preventing a substantial percentage of the solar radiation from reaching the surface. An almost palpable haze allowed sunspots to be viewed with the naked eye on a rather watery Sun, while spectacular sunsets caused by volcanic dust and aerosols in the stratosphere may have provided the inspiration for artist Joseph Turner's most colorful works.

As the northern hemisphere was plunged into winter the after-effects of the eruption became less of a curiosity and more a matter for survival. In Europe, Spring 1816 brought torrential rains and chilly temperatures that resulted in poor crop yields, high grain costs, and food riots in France and Britain. Winter wheat harvests failed across mainland Europe and, in Ireland, famine and an accompanying typhus epidemic is estimated to have killed over 50,000

LEFT The Tambora eruption caused a dramatic fall in northern hemisphere temperatures—reflected in this chart for North America—and played havoc with the weather. OPPOSITE Famine, war, and death stalked Europe following the appalling weather and failed harvests.

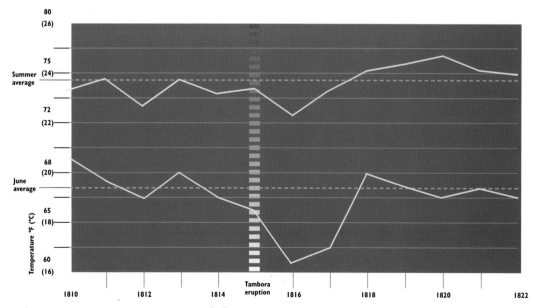

people over the next two years. Famine also stalked landlocked Switzerland, where in the high mountain passes bands of starving bandits stole Russian wheat being brought north from Italian ports.

Snowfall in June

North America fared little better, with heavy snow falling over much of northeastern United States in June, and more freezing weather and snow in July, August, and September. The unprecedented, bitterly cold weather destroyed most of the corn—a staple crop of the region —and ravaged the harvests of beans, wheat, squash, and other less hardy crops. By Winter 1816, shortages pushed up grain prices to eight times their normal prices. Unable to afford hay and corn to feed to their livestock, farmers fed them mackerel instead, and supplemented their diets with raccoons, groundhogs, and wild plants. The anger that many people felt was taken out on politicians, and the elections of 1816 saw more congressmen lose office than any other time since. Other people voted with their feet, abandoning farms across the northeast for the perceived "promised land" of the Wild West.

The freezing weather of 1816 was the result of a temperature fall of 1.3°F (0.7°C) all over the globe, caused by the veil of Tambora aerosols draped across the planet. At the latitudes of Europe and North America, however, the temperature fall may have been as great as 3°F (2°C), reflecting the fact that here the Sun's rays strike the Earth at a more acute angle and so must pass through a much thicker aerosol veil than at the tropics. It even appears that the gloom of the year seeped into the literary world—the cold, wet conditions set the mood for Mary Shelley to pen the words of the Gothic horror *Frankenstein*, as she spent the summer housebound by Lake Geneva.

A flaming future

>>>There is now little doubt that the Earth is heating up as a consequence of the polluting activities of our modern, industrial society.

>Forecasts for the planet in 2100 do vary, but all predict a much warmer world threatened by rising sea levels. Worst-case scenarios visualize blistering average temperatures almost 10°F (6°C) higher than today, with sea levels over 30 inches (80cm) higher resulting in flooding of many low-lying areas.

We are now familiar with the ways in which volcanic activity can affect and modify the climate, but can the reverse ever happen? Can climate change trigger volcanic eruptions? Wild as this may sound, there is now evidence that it may be possible. In fact, the Earth's climate and its volcanoes may be locked into a cause-and-effect feedback loop, with volcanoes triggering climate change and climate change causing volcanoes to erupt. The problem is, which comes first? Which is the chicken and which the egg?

The most dramatic climate change in our planet's long history can be found during the Ice Ages, when rapid and major temperature swings led to repeated expansions and contractions of the polar ice caps. These climatic oscillations occurred four times during the last Ice Age and lasted for around 2 million years, ending just 10,000 years ago. The same period was also characterized by dramatically increased levels of volcanic activity, supporting a link between the two. But just how can changing temperatures cause volcanoes to erupt? Well, the answer is, they can't—the real link appears to be not between volcanoes and temperature changes but between volcanoes and the large, rapid changes in sea levels that accompany them. Between the warmest interglacial and the frigid depths of a glacial period, average global sea levels change by about 400 feet (120m). This involves a huge redistribution of water across the planet and establishes conditions that are particularly favorable for volcanoes to erupt.

The stress on our planet

On a global basis, moving such huge volumes of water around can actually affect the rate at which the Earth spins on its axis. Increasing or reducing this spin-rate places stress on the Earth's crust that can trigger the ascent of magma and its eruption. On a regional scale, a large sea level change has a similar effect on crustal stresses. For example, a sea level fall of 400 feet (120m) would dramatically reduce the loading of the continental margins, permitting stored magma to erupt more readily. At the same time, increased numbers of earthquakes generated as the crust adapts to the new stress regime could induce collapses of volcanoes, triggering eruptions in the manner of Mount St. Helens. At individual coastal or island volcanoes, dramatic sea level changes may also promote eruptions in a number of ways—including erosion leading to collapse-induced eruptions, modifications of the water table increasing the chances of magma and water combining, and the opening of new fractures and faults.

Recent research has revealed a link between the timing of volcanic eruptions over the last 100,000 years and the rate of sea level change, suggesting that the more rapidly sea level goes up and down the more volcanic eruptions there are. As over 60 percent of all volcanoes are either coastal or form islands, this relationship is not surprising. Some of the rates of sea level change over the past ten millennia have been extremely impressive, perhaps as high as 30 feet (10m) in 150 years. The big question now is: will the sea level rises due to global warming be sufficient to cause an increase in volcanic activity and further add to our woes? It is most unlikely that the 30 inch (80cm) worst-case forecast will

be sufficient, but the ultimate consequence of unabated and unmitigated global warming will be catastrophic melting of the polar ice caps and rises of more than 230–260 feet (70–80m). It would be no surprise if such a huge rise led to an unprecedented burst of renewed volcanic activity, resulting in a flaming future for us all.

ABOVE If sea levels rise fast enough, many more coastal and island volcanoes may burst into life.

BELOW Changing sea levels can trigger volcanic eruptions in a multitude of different ways.

At individual volcanoes
● Rising water tables
● Increased erosion
● Activation of faults
● Debuttressing of faults
● Reduced pressures
● Opening of new fractures

At a regional and global level
● Crustal deformation due to redistribution of planetary water
● Changes in the Earth's spin

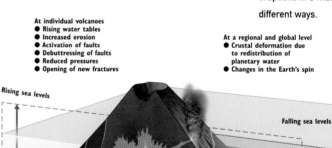

Rising sea levels

Falling sea levels

Magma chamber

Increased earthquakes

Stress changes

NOWHERE TO RUN TO, NOWHERE TO HIDE

Spectacular and awe-inspiring as recent volcanic blasts seem, they are nothing but ineffectual sparks compared to the blazing fireworks Vulcan has in store for the human race in years to come.

The size of past eruptions can be estimated through observations of two products. First, the volume and extent of ejected deposits and, second, the size of the hole—known as a caldera—excavated by the blast. Looking back in time—sometimes only a few thousand years—the scale of both is truly frightening.

Crater Lake represents the water-filled cavity left when Mount Mazama blew itself apart barely 7,000 years ago devastating much of what is now Oregon State (U.S.). Heading northeast from Crater Lake, the steaming geothermal fields of Yellowstone National Park (Wyoming, U.S.) lie within three gigantic calderas formed by some of the greatest eruptions ever, all within the last 2 million years or so. On the other side of the Pacific, in New Zealand, the spectacular Lake Taupo was formed by a huge eruption 26,500 years ago that poured out ash and pyroclastic flows across most of the North Island.

But even more terrifying volcanic events lie in store, with the power to dramatically modify the global climate and plunge the world into a dark and freezing volcanic winter. This last happened 73,500 years ago—no time at all in geological terms—when a great mass of magma blasted through the crust at Toba on the Indonesian island of Sumatra. As a result the skies darkened and the Sun was blotted out across the planet—perhaps for years. Digging deeper into the depths of Earth history reveals an even greater horror: huge floods of basalt lava spewing unstoppably across areas as large as a continent and pumping trillions of tons of climate-cooling sulfur gases into the atmosphere.

Mind-boggling as such events might seem, we would be wrong to imagine that they are irrelevant curiosities from the Earth's distant past. Such apparently unprecedented geophysical phenomena are a normal part of our planet's life cycle. They will happen again and when they do we will all suffer the awful consequences—no matter where we live.

OPPOSITE The hot springs, steaming pools, and geysers of Yellowstone Park in Wyoming testify to the magma that still lurks not far below the surface. BELOW The huge, lake-filled cavity known as Crater Lake was formed when Oregon's Mount Mazama blew itself apart around 7,000 years ago.

The biggest bangs of all

>>>As seismologists use the Richter Scale to provide a measure of the size or magnitude of an earthquake, so volcanologists use the more mundanely labeled Volcanic Explosivity Index (VEI) to describe the scale of a volcanic eruption.

>While seismologists are able to determine the energy of a quake directly, using a seismograph, volcanologists have no single measure that can provide them with a reliable estimate of the amount of energy a volcanic eruption releases. It is, therefore, very difficult to determine the true size of an eruption and many qualitative estimates that have been made in the past have proved to be wildly inaccurate. This is primarily because to an individual, and particularly one who has never before experienced an eruption, any estimate of size is liable to be subjective. An inhabitant of the Caribbean island of Montserrat, for example, might feel justified in expounding on the subject of the huge eruption that continues to devastate the island; but to a survivor of the 1991 Pinatubo eruption in the Philippines, the recent rumbles of Montserrat's Soufrière Hills volcano would represent little more than the bangs and pops of an innocuous firecracker.

The VEI was devised in the early 1980s by volcanologists Chris Newhall and Steve Self to address problems of comparability between eruptions and to provide a more quantitative measure of their size. The index uses a number of measurable parameters to provide a clue to the size of an eruption, including the volume of explosively ejected material and the height to which it is ejected. Like the Richter Scale, the VEI is open-ended, although so far no evidence has come to light for any eruption registering more than an 8. The VEI is also logarithmic—again following in the footsteps of Richter—so each point on the scale represents an eruption ten

times larger than the one below.

Right at the base of the index, VEI 0 is reserved for non-explosive eruptions: quiet oozings of low-viscosity basaltic lava common on the volcanoes of Hawaii and Iceland. Eruptions that register at 1 and 2 on the scale are somewhat more violent, ejecting sufficient ash to perhaps cover New York or London in a light dusting. Things liven up a little more, however, at points 3, 4, and 5. These larger values designate the progressively more violent explosive eruptions that are characteristic of the stickier, gas-rich andesitic and rhyolitic magmas of subduction zone volcanoes. The twin eruptions of Tavurvur and Vulcan at Rabaul (Papua New Guinea) in 1994 scored a 4 on the index, while

BELOW The incredibly violent eruption of the Taupo caldera volcano in New Zealand in 186 A.D. scored a VEI of 6 and covered much of the North Island with pyroclastic flow deposits tens of feet thick.

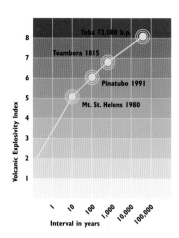

LEFT The larger the eruption, the less frequently it occurs. Eruptions on a scale of Mount St. Helens occur every decade, but we have to wait 50,000 years or so between super-eruptions.

RIGHT The twin eruptions in Papua New Guinea of Rabaul's Tavurvur and Vulcan volcanoes achieved a 4 on the VEI.

Volcanic Explosivity Index (VEI)

VEI	VOLUME OF TEPHRA (M³)	HEIGHT OF ERUPTION COLUMN (KM)	ERUPTION RATE (KG/SECOND)	DURATION OF BLAST (HOURS)	GENERAL DESCRIPTION	EXAMPLE
0	< 10,000	< 0.1	100 TO 1,000	<1	NON-EXPLOSIVE	KILAUEA, HAWAII
1	10,000 TO 1 MILLION	0.1 to 1	1,000 TO 10,000	<1	SMALL	STROMBOLI, ITALY
2	1 MILLION TO 10 MILLION	1 to 5	10,000 TO 100,000	1 TO 6	MODERATE	GALERAS, COLOMBIA
3	10 MILLION TO 100 MILLION	3 to 15	100,000 TO 1 MILLION	1 TO 12	MODERATE TO LARGE	MONTSERRAT, CARRIBEAN
4	100 MILLION TO 1 BILLION	10 to 25	1 MILLION TO 10 MILLION	6 TO 12	LARGE	RABAUL, PNG
5	1 BILLION TO 10 BILLION	>25	10 MILLION TO 100 MILLION	>12	VERY LARGE	MOUNT ST. HELENS, U.S.
6	10 BILLION TO 100 BILLION	>25	100 MILLION TO 1 BILLION	>12	VERY LARGE	PINATUBO, PHILIPPINES
7	100 BILLION TO 1 TRILLION	>25	>1 BILLION	>12	EXTREMELY LARGE	TAMBORA, INDONESIA
8	>1 TRILLION	>25	>1 BILLION	>12	SUPER-ERUPTION	YELLOWSTONE, U.S.

the spectacular 1980 Mount St. Helens blast registered at 5. It is at point 6 on the index, however, that things start to get really serious, and eruptions on this scale and above have impacts that may extend far from the volcano. Both the 1991 eruption of Pinatubo, and the great Krakatoa blast of 1883 were designated VEI 6, making them ten times larger than Mount St. Helens. Both caused regional devastation and pumped sufficient gas into the atmosphere to cause a noticeable cooling across the planet. On an altogether more impressive scale, the VEI 7 eruption of Tambora (Indonesia) in 1815 led to the deaths, directly or indirectly, of nearly 100,000 people and cast a pall of gas across Europe and North America that led to the year without a summer. The volume of debris ejected by the Tambora blast is estimated at over 40 square miles (100 sq.km). Point 8 on the VEI is reserved for the biggest bangs of all; super-eruptions that—fortunately for us—occur very infrequently. The last super-eruption at Toba, 73,500 years ago blasted out enough ash and debris to bury the whole of greater London to a depth of two-thirds of a mile (1km).

Restless giants

>>> The legacy of the Toba super-eruption is a spectacular lake-filled caldera over 60 miles (100km) long, which is almost matched in size—at 50 miles (80km)—by the largest of the three, giant Yellowstone calderas.

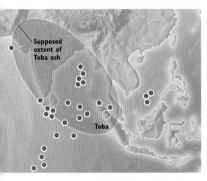

ABOVE Toba ash has been discovered in deep sea cores across southeast Asia—far beyond its original supposed extent.

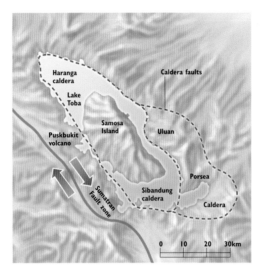

ABOVE The island and much of the Toba lake margin has been uplifted by up to 1,300 feet (400m) indicating that fresh magma is still pushing up from below.

LEFT The 60-mile (100-km) long Toba caldera is now filled with a lake and has become a well-known tourist resort.

> Other great volcanic blasts have also excavated calderas at Long Valley in California, Campi Flegrei in the Italian bay of Naples, Rabaul in Papua New Guinea, and Taupo in New Zealand.

Although each of these volcanoes has destroyed itself in a past eruption, it would be wrong and dangerous to conclude that they are now extinct. Far from it; all of these calderas are described by volcanologists as restless and merit careful watching. At each of them magma stirs not far beneath the surface generating ground tremors and causing the ground to rise and fall.

At Campi Flegrei near Naples, the ground swelled over a huge area by almost 7 feet (2m) during the 1970s and 1980s, causing consternation among the millions of people who live close by and desperately increasing fears of an imminent eruption. The situation has now returned to normal but, with the last eruption occurring just over 400 years ago, nobody can afford to relax. The residents of Naples are only too aware that similar restlessness at Rabaul caldera in Papua New Guinea during the 1980s was followed by a devastating eruption in 1994.

In the United States, there is increasing concern over signs of life at the huge Long Valley caldera in California. Numerous earthquakes, swelling of the ground surface, and the release of carbon dioxide gas since 1980 seem to point to new magma approaching the surface. Activity is focused beneath the famous Mammoth Mountain ski resort, and both scientists and locals are wondering what the future will hold.

Further north in Wyoming, the Yellowstone caldera is also far from dead. As at other restless calderas, the ground rises and subsides periodically while earthquakes—some large enough to damage property and take lives—regularly shake the region. Most significantly, hot magma not far beneath the surface heats up rainwater percolating into the ground, and sends it back again in the form of bubbling mud pots, steaming pools, and spectacular geysers.

Certain obliteration

When one of these restless giants awakens, what can we expect? Obliteration! The last Yellowstone super-eruption, which occurred around 630,000 years ago, sent blistering pyroclastic flows across what is now Wyoming and neighboring states, sufficient to bury the entire country in a deposit 3 inches (8cm) deep. Ash poured down from the skies over more than half of the country, falling as far as El Paso in Texas, and Los Angeles in California. A similar blast today would paralyze the United States and bring the economy to its knees. The climatic impact of the eruption, together with the resulting economic effects, would plunge the planet into years of mayhem and anarchy that would see our global society fighting to survive.

What volcanologists fear most is a future eruption on the scale of the cataclysmic Toba blast. Ash from the last Toba eruption is found in deep sea sediments all over south and Southeast Asia and recent estimates suggest that the volume of material ejected might have been as much 1,400 cubic miles (6,000cu.km)—over twice as big as the Yellowstone blast. Like its U.S. counterpart, the Toba caldera continues to swell and shiver, revealing that magma still churns beneath the surface. Lake sediments deposited after the eruption have been thrust upward by over 1,300 feet (400m) and a new island has grown in the lake center in response to pressure from below. Toba will almost certainly erupt again but no one knows when.

RIGHT Super-eruptions in Yellowstone National Park have excavated three huge calderas over the past 2 million years.
BELOW Calderas form when magma is explosively evacuated from circular fractures and the central block subsides into the resulting cavity.

YELLOWSTONE NATIONAL PARK

Mammoth

Tower Junction

Norris Junction

Canyon Junction

Madison Junction

West Yellowstone

Youngest Yellowstone caldera 630,000 years old

Elephant Back fault zone

Lake Junction

Old Faithful

Yellowstone Lake

West Thumb

Second Yellowstone caldera (Henry's Fork or Island Park caldera) 1.3 million years old

First Yellowstone caldera 2 million years old

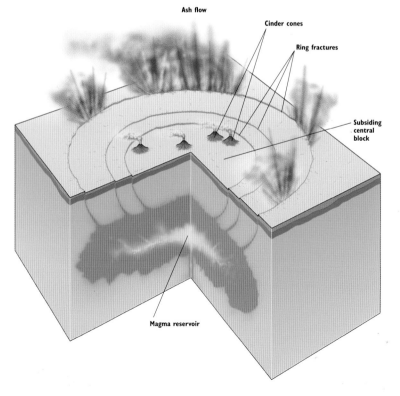

Ash flow

Cinder cones

Ring fractures

Subsiding central block

Magma reservoir

Floods of doom

>>> Although terrifying in their violence, explosive super-eruptions on the scale of Toba are far from the greatest volcanic cataclysms known.

> Such earth-shattering blasts involve viscous, rhyolitic magma with a high gas content that rips the magma apart on eruption. Going back tens or hundreds of millions of years into Earth's history reveals, however, that the largest eruptions of all involve the relatively quiet extrusion of low viscosity basalt magma. These are no ordinary lava flows, however, but huge floods of molten rock of biblical proportions. Known as flood basalts, these spectacular outpourings have been identified in many parts of the world, including the northwest United States, India, southern Africa and even northwest Scotland, and demonstrate a range of ages.

By far the greatest flood of basalt lava, however, breached the surface 248 million years ago in what is now the frozen wasteland of northern Siberia. Although the numbers remain a matter for conjecture, it looks as if the total volume of lava erupted amounted to 700,000 cubic miles (3 million cu.km)—500 times the volume of Toba. Recent research suggests that the lavas poured out incredibly rapidly, covering an area of over 10 million square miles (25 million sq.km)—equivalent to three times the area of the U.S.—in perhaps only a few hundred thousand years. So much lava poured forth, in fact, that if spread out evenly it would cover the

BELOW Enormous floods of low viscosity basalt magma are found in many parts of the world and have been implicated in mass extinctions within the geological record, including the obliteration of the dinosaurs.

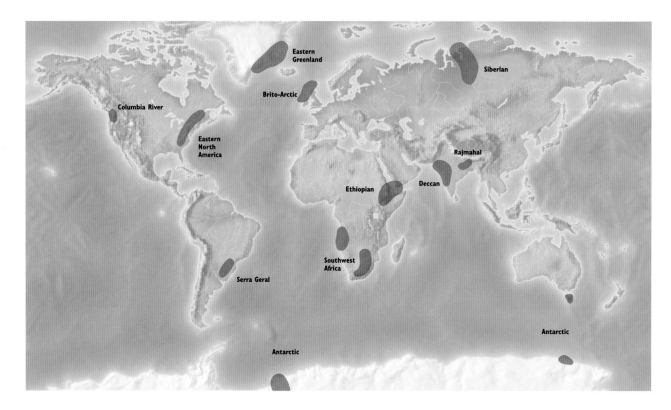

Labels on map: Eastern Greenland, Siberian, Brito-Arctic, Columbia River, Eastern North America, Rajmahal, Deccan, Ethiopian, Southwest Africa, Serra Geral, Antarctic, Antarctic

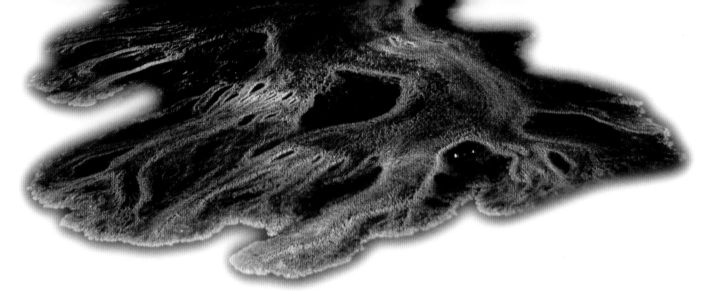

entire planet to a depth of almost 10 feet (3m). The impact of these lava floods on the planet's ecosystems was utterly lethal.

Prior to the volcanic cataclysm, during the geological period known as the Permian, the Earth teemed with life. In contrast, the world during Triassic times, which followed the end of the eruption, was a barren and empty place. It seems that the extrusion of the Siberian Traps— as the great flows are known—came very close to wiping life from the face of the Earth. Fifty-seven percent of all families and 95 percent of all species disappeared in the greatest mass extinction of all time. Eighty million years later life had recovered, but many families would never be seen again—most notably the trilobites, armored marine invertebrates well-known to all avid fossil hunters.

What really killed the dinosaurs?

Some scientists think that a second mass extinction event 65 million years ago was also the result of a great outpouring of basalt lava, this time in the Deccan region of northwest India. One-third the volume of the Siberian lavas, these Deccan Traps are up to a mile (2km) thick and, again, coincide with the loss of many species, including the dinosaurs. There is now, however, clear evidence that a comet hit the Earth at this time and many believe that this was the reason for the demise of the dinosaurs and other species. Some volcanologists, at least, beg to differ but the jury currently remains out.

In any case, just how can a volcanic eruption —however large—almost wipe out life across the whole planet? The answer, it seems, lies not in the lavas themselves but, once again, in the accompanying release of gas. In the short term, sulfur gases can lead to severe cooling. However, over a long period of time, perhaps many thousands of years, carbon dioxide could prove to be the real killer. As we are currently realizing at our expense, carbon dioxide is a greenhouse gas, meaning that it reduces the amount of heat that the Earth radiates back into space, thus warming the planet up. Some scientists argue that the volume of carbon dioxide pumped into the atmosphere during a flood basalt eruption would dwarf the rise in emissions caused by our increasingly polluting society. This, they believe, would lead to a catastrophic rise in global temperatures and the widespread deaths of many species that could not adapt to the hotter conditions. Perhaps flood basalts represent a warning from Mother Nature—one we should take much more seriously if we do not wish to follow in the dinosaurs' footsteps.

ABOVE Floods of lava, some large enough to bury entire continents, have occurred throughout Earth's long history.

Volcanic winter

>>> The short-term cooling effect of explosive volcanic eruptions is now well established.

> Recognized following Tambora and Krakatoa in the nineteenth century, after the 1982 eruption of El Chichón in Mexico, and the 1991 Pinatubo blast in the Philippines, post-eruption cooling amounted to less than 2°F (1°C) and persisted for only a year or two. Following a future super-eruption, the cooling would be far more dramatic.

Bring on the Ice Age

As well as ejecting 1,450 cubic miles (6,000 cu.km) of ash and debris, Toba—which erupted 73,500 years ago—also appears to have pumped enough sulfur dioxide gas into the stratosphere to mix with atmospheric water and form 5 billion tons of sulfuric acid aerosols. Spreading out across the planet, these droplets would have rapidly absorbed the Sun's rays or reflected them back into space, leading to a dramatic cooling at the Earth's surface. Calculations suggest that the fall in global temperatures after the Toba eruption may have been between 5 and 8°F (3–5°C), and up to 25°F (15°C) in some regions. Such a drop may have been sufficient to trigger a volcanic winter.

Studies from boreholes drilled deep into the Greenland ice sheet also reveal that the Toba eruption was followed by a 1,000 year long "cold snap," which heralded the last Ice Age —a time of terrible cold from which the planet only emerged around 10,000 years ago. The Earth was already getting colder when Toba went off, but it is feasible that the succeeding volcanic winter accelerated the planet's slide toward full Ice Age conditions.

Given that a super-eruption will certainly happen again, the big question is—can the human race survive? Our prospects don't look too rosy. A worse-case scenario like Toba suggests that the eruption may have blasted sufficient sulfur gases into the stratosphere to cut the amount of sunlight reaching the surface of the Earth by over 99 percent, plunging the world into perpetual night, perhaps for several years.

Photosynthesis is likely to have slowed to almost nothing, leading to a failure of harvests across the planet. As even developed countries, such as the United Kingdom and the United States have—at any one time—little more than four weeks of food supplies available to their populations, the prospects of surviving for years with little or no replenishment are grim. Some estimate that over 1 billion will die in the aftermath of a super-eruption, with our modern society being knocked back to the Dark Ages.

We can get an inkling of what we might face by examining the impact the Toba eruption may

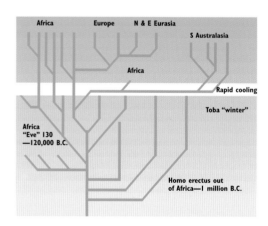

ABOVE The bitter volcanic winter that followed the giant Toba eruption may have wiped out many branches of humanity.

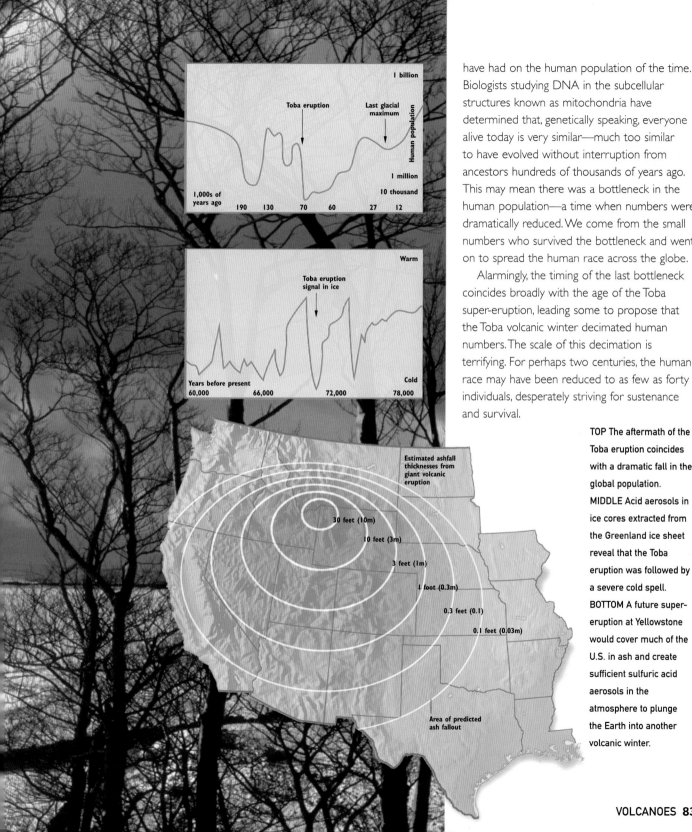

have had on the human population of the time. Biologists studying DNA in the subcellular structures known as mitochondria have determined that, genetically speaking, everyone alive today is very similar—much too similar to have evolved without interruption from ancestors hundreds of thousands of years ago. This may mean there was a bottleneck in the human population—a time when numbers were dramatically reduced. We come from the small numbers who survived the bottleneck and went on to spread the human race across the globe.

Alarmingly, the timing of the last bottleneck coincides broadly with the age of the Toba super-eruption, leading some to propose that the Toba volcanic winter decimated human numbers. The scale of this decimation is terrifying. For perhaps two centuries, the human race may have been reduced to as few as forty individuals, desperately striving for sustenance and survival.

TOP The aftermath of the Toba eruption coincides with a dramatic fall in the global population.
MIDDLE Acid aerosols in ice cores extracted from the Greenland ice sheet reveal that the Toba eruption was followed by a severe cold spell.
BOTTOM A future super-eruption at Yellowstone would cover much of the U.S. in ash and create sufficient sulfuric acid aerosols in the atmosphere to plunge the Earth into another volcanic winter.

Top chart labels: 1 billion; Toba eruption; Last glacial maximum; Human population; 1 million; 10 thousand; 1,000s of years ago; 190 130 70 60 27 12

Middle chart labels: Warm; Toba eruption signal in ice; Cold; Years before present; 60,000 66,000 72,000 78,000

Bottom map labels: Estimated ashfall thicknesses from giant volcanic eruption; 30 feet (10m); 10 feet (3m); 3 feet (1m); 1 foot (0.3m); 0.3 feet (0.1); 0.1 feet (0.03m); Area of predicted ash fallout

THE TIDE OF DEATH

The largest volcanoes on Earth are not found on the continents, but are built up from the deep ocean floor. Volcanic oceanic islands that make up the Hawaii and Canary archipelagoes have their roots in the oceanic crust from which they rise to extraordinary heights.

The Big Island of Hawaii is made up largely from the giant Mauna Loa volcano that towers 6 miles (10km) from the black depths of the Pacific Ocean, making it taller than Mount Everest. As active ocean island volcanoes continue to grow so they also become more and more unstable. As part of their normal life cycles they cope with this instability by periodically sloughing off huge masses of rock in the form of some of the greatest landslides on the planet. Some of the Hawaiian landslides have volumes in excess of 240 cubic miles (1,000 cubic km) and traveled for thousands of miles across the seabed. Over the last few decades, both manned and robot submersibles have photographed the seabed around the Hawaiian islands. These photographs have revealed the entire chain to be sitting on a massive apron of debris that has been periodically divested by its constituent volcanoes over tens of millions of years—over 70 great landslides in all.

It is difficult to estimate just how often these huge collapses happen, but on average they seem to occur once every 25,000 to 100,000 years. Such spectacular events should not be regarded merely as scientific curiosities, because the impact of a future collapse on human society may be cataclysmic. Evidence from past collapses and from computer simulation of future events suggests that as such gigantic landslides enter the ocean they generate enormous tsunami—

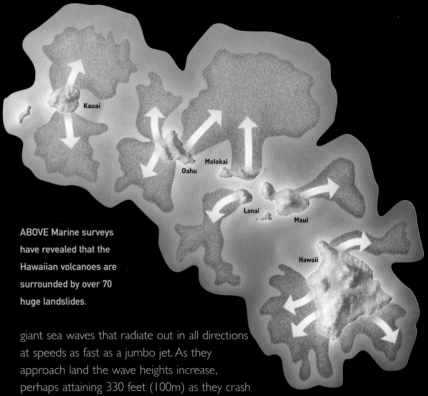

ABOVE Marine surveys have revealed that the Hawaiian volcanoes are surrounded by over 70 huge landslides.

giant sea waves that radiate out in all directions at speeds as fast as a jumbo jet. As they approach land the wave heights increase, perhaps attaining 330 feet (100m) as they crash into a coastline. Tsunami generated by the next Hawaiian landslide will batter all the Pacific Rim countries within 12 hours, killing millions and obliterating some of the world's greatest cities. Nor is the threat confined only to the Pacific: in the north Atlantic Ocean, a volcano on the Canary Island of La Palma is becoming increasingly and alarmingly unstable.

OPPOSITE In addition to damaging lava flows, the Hawaiian island volcanoes undergo periodic collapse, generating massive landslides into the sea that send tsunami hurtling across the Pacific Ocean.

Countdown to catastrophe

>>>Belying its tranquil setting, the spectacular Canary Island of La Palma represents a disaster waiting to happen, and one on a scale unprecedented in modern times.

>The island is entirely volcanic and is no stranger to great landslide events. Half a million years ago, much of the great Taburiente volcano, which was active in the northern part of the island, crashed into the sea, leaving behind a gigantic cliff-bounded amphitheater 6 miles (10km) across. More recently, and further south, a new upstart volcano—known as the Cumbre Vieja—has grown rapidly over the past 125,000 years. It towers 1.2 miles (2km) above sea level and an even more impressive 4 miles (6km) above the bed of the north Atlantic—three-quarters the height of Everest.

The Cumbre Vieja differs markedly from the classic cone-shaped structures reminiscent of Japan's Mount Fuji: it consists of a 12-mile (20-km) ridge that is little more than 9 miles (15km) across at its widest. The flanks of the volcano are steep sided, plunging into the sea at angles sometimes in excess of 30°. Such steep slopes rarely last long before succumbing to gravity and causing landslides.

On the verge of collapse

The steep slopes of volcanoes are even more unstable than their counterparts elsewhere and a number of factors—in addition to gravity—conspire to trigger their collapse. While fresh magma pushing up into a volcano can bulge out the flanks, ground-shaking due to earthquakes associated with the rise and eruption of magma can cause the bulging flank to break off and slide down-slope, as happened during the 1980 eruption of Mount St. Helens. Furthermore, the same fresh magma can heat up water contained in voids and cavities in the rock, causing it to expand and exert pressure on its surroundings. If the slope is teetering on the edge of instability, this extra push can be sufficient to send a huge landslide hurtling downhill.

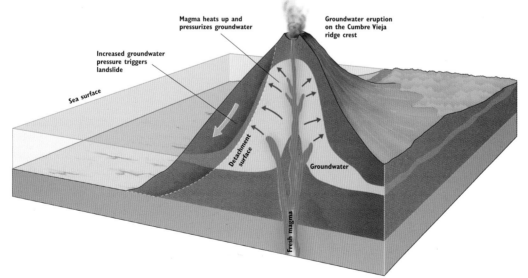

Magma heats up and pressurizes groundwater

Groundwater eruption on the Cumbre Vieja ridge crest

Increased groundwater pressure triggers landslide

Sea surface

Detachment surface

Groundwater

Fresh magma

LEFT The west flank of Cumbre Vieja will collapse during a future eruption, as fresh magma triggers a gigantic landslide.

The Cumbre Vieja is one of the fastest growing volcanoes on the planet and, although quiet at the moment, it is far from extinct. Twice during the twentieth century—in 1949 and 1971—fresh magma blasted forth, generating lava flows and ash. The most recent eruption occurred in the far south of the island and posed little threat to any but a few local inhabitants and none to the long-term stability of the volcano. In this latter respect, however, the 1949 eruption was very different.

After more than two centuries of tranquility, the year saw a violent arrival of magma at the surface with ash hurled skyward and lava pouring down the flanks. During the course of the eruption, strong earthquakes accompanied the tearing open of the ground and the formation of a 2-mile (3-km) long system of fractures along the volcano's summit ridge. When the Spanish seismologist Bonelli Rubio visited the eruption site in early July he discovered that everything to the west of the fractures had dropped seaward by up to 13 feet (4m).

Scientists now think that the 1949 eruption caused the detachment of much of the west flank of the volcano, making it the youngest giant volcanic landslide in the world. Estimates suggest that a chunk of rock the size of greater London —perhaps 48 cubic miles (200 cu.km) in all —is now poised precariously above the north Atlantic Ocean, waiting to plummet seaward at the slightest provocation. Recent measurements made using the Global Positioning System suggest that the entire mass may be creeping downslope at half an inch (1cm) or so a year. While it is unlikely that a major collapse will be triggered without the impetus of an eruption and the emplacement of fresh, hot magma, two big questions remain: will the collapse occur in ten years or in 10,000 years? And what will be the implications of the arrival of an enormous mass of volcanic rock in the North Atlantic?

ABOVE The Cumbre Vieja volcano is one of the world's fastest growing volcanoes.

TOP Monitoring of the volcano suggests that the west flank may be continuing to creep very slowly seaward. BOTTOM During the 1949 eruption, the western flank of the Cumbre Vieja dropped about 4 yards or meters toward the sea along a network of open fractures, and is now poised precariously over the North Atlantic.

Wipe-out!

>>> It is inevitable that dumping a gigantic chunk of rock into the ocean will generate large waves, but few have any idea of just how large, how extensive, and how destructive these waves can be.

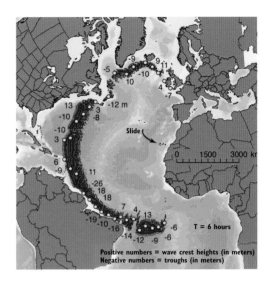

> In particular, because such waves have never been experienced in modern times, many people—including some scientists—find it difficult to believe the extent of destruction possible. When the western flank of the Cumbre Vieja eventually collapses, the tsunami that will hurtle out from La Palma will obliterate towns and cities all around the rim of the Atlantic Ocean; the Caribbean and the eastern coast of the United States will take the brunt of the battering. There is mounting evidence for the existence and scale of ancient tsunami generated by collapses of the Canary Island volcanoes and it makes very scary reading.

Tsunami culprit

South of La Palma lies neighboring El Hierro. Also volcanically active, this island owes its shape to the sculpting effects of three enormous collapses that have formed three great bays. One of these—El Golfo—is a spectacular semi-amphitheater some 12 miles (20km) across and bounded on the landward side by precipitous cliffs two-thirds of a mile (1km) high. Clearly, a huge mass of material has been displaced from El Golfo and photographic imagery of the adjacent seafloor reveals just where it has gone. Stretching 50 miles (80km) from the coastline is a massive landslide containing individual blocks of rock two-thirds of a mile (1km) across with an estimated volume of around 24 cubic miles (100 cu.km). The age of the collapse appears to lie between 90,000 and 130,000 years. Ominously, this is comparable with the age of some enigmatic deposits that have recently

come to light 1,550 miles (2,500km) away on the other side of the Atlantic.

On the Bahamian island of Eleuthera coral limestone bolders the size of houses, weighing thousands of tons, have been stranded 65 feet (20m) above sea level and up to 1,600 feet (500m) inland. It is difficult to imagine how these huge chunks of rock could have reached their current position without being catapulted there by gigantic waves. Furthermore, all along the Bahamian archipelago, the same waves—clearly coming from the direction of the Canary Islands—have piled sand into huge wedges several miles long that punch through gaps between the islands.

Although it has been suggested that the Bahamian features were the cause of unprecedented storm waves, they may, in fact, be the first evidence on the western side of the Atlantic of giant tsunami formed by the collapse of a Canary Island volcano. Signs are also being uncovered in the Canaries themselves, and tsunami deposits of cobbles and seashells have recently been discovered on the neighboring islands of Gran Canaria and Fuerteventura, up to 330 feet (100m) above current sea level and far

ABOVE The collapse of the Cumbre Vieja volcano will send huge tsunami hurtling across the Atlantic. After six hours, waves tens of meters high will be bearing down on the eastern coast of the United States.

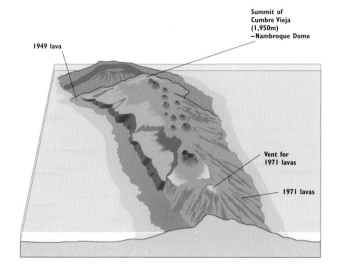

1949 lava

Summit of
Cumbre Vieja
(1,950m)
– Nambroque Dome

Vent for
1971 lavas

1971 lavas

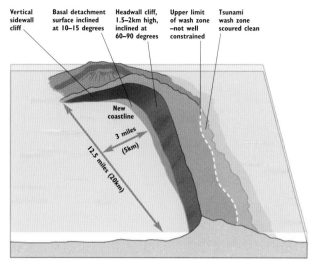

Vertical
sidewall
cliff

Basal detachment
surface inclined
at 10–15 degrees

Headwall cliff,
1.5–2km high,
inclined at
60–90 degrees

Upper limit
of wash zone
–not well
constrained

Tsunami
wash zone
scoured clean

New
coastline

3 miles
(5km)

12.5 miles (20km)

BELOW Were these gigantic blocks of coral dumped on land by volcanic tsunami that crashed into the Bahamas some 100.000 years ago?

ABOVE Before and after: the future collapse of the Cumbre Vieja will remove much of the western half of the Canary Island of La Palma.

inland. It is difficult to visualize the devastation that would ensue today from the impact of such enormous waves, but the La Palma threat makes it vital that we know what we are up against.

Once the west flank of the Cumbre Vieja volcano slides into the sea, mass destruction and loss of life are inevitable. Within the first hour or two, huge tsunami will surge up the flanks of all the Canary Islands and obliterate coastal settlements, including Santa Cruz, La Palma's capital. An hour later, the waves will have crashed against the north African coast and blasted through the Gibraltar Straits, trashing Casablanca, Tangier, and Cadiz. Traveling faster than a jet plane, the tsunami would shortly be battering the coasts of the southern United Kingdom, but the worst would be reserved for the countries of the western Atlantic. Barely 6.5 hours after the collapse the Bahamas will be scoured clean of all human life, while an hour or so later wave after wave—each perhaps 160 feet (50m) or more in height—will crash into the cities of the U.S. East Coast. Given only eight hours or so warning, it is difficult to see how casualty numbers totaling in the millions can be avoided.

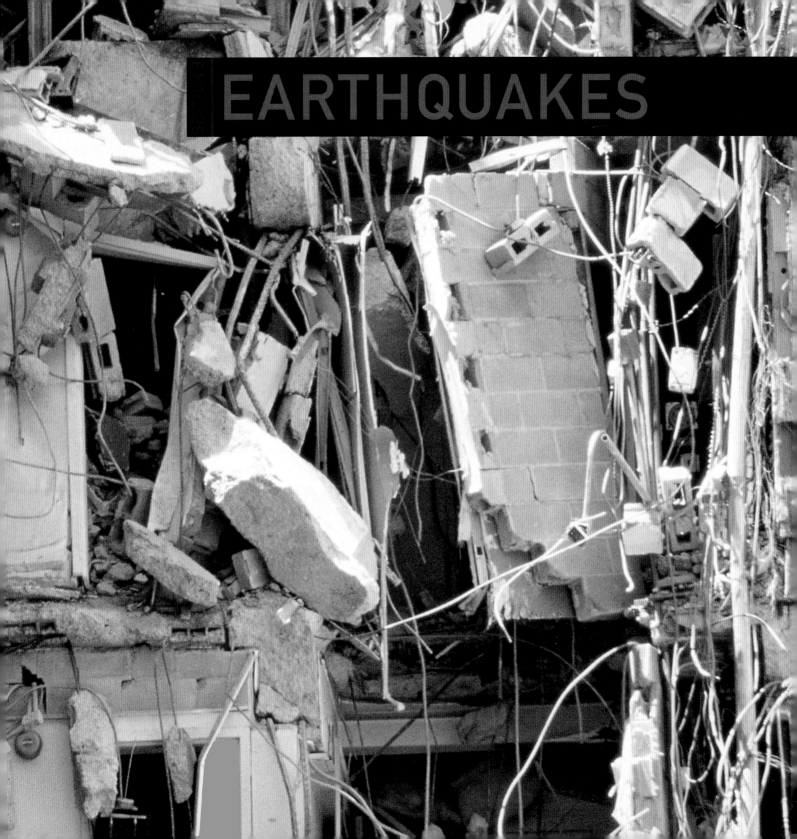

EARTHQUAKES

DID THE EARTH MOVE FOR YOU?

Throughout our planet's long and tempestuous history, earthquakes have always been with us.

However, only since our descendants began to build permanent homes and congregate in larger settlements, have they really started to seriously affect our lives. There is no question that earthquake engineers have hit the nail squarely on the head with their mantra "it's buildings, not earthquakes, that kill people." Furthermore, the heavier the building, the more likely we are to die if it collapses on us. If all those who lived in quake-prone regions dwelled in lightly built wooden structures then the death tolls from collapsing buildings would be dramatically reduced—though, unfortunately, many more people would probably be killed in the subsequent raging fires that would soon reduce such structures to ashes.

The answer to minimizing injury, loss of life, and property damage during an earthquake then is not just wooden buildings. With successful earthquake prediction as far off as ever, mitigating their awful impact must be a combination of appropriate building construction, enforcement of building codes, and community preparedness. The 1989 Loma Prieta and 1994 Northridge earthquakes in California clearly demonstrated how well-constructed buildings and strictly enforced building standards can dramatically reduce the impact and casualty figures of even large quakes. In sharp contrast, the destruction and huge death toll arising from the 1999 Izmit earthquake in Turkey highlighted the terrible consequences of inappropriate and poorly enforced construction in a zone of high

seismic risk. The effect of poor preparedness was also made obvious to all who surveyed the aftermath of the 1995 Kobe earthquake in Japan, where survivors had to fend for themselves for up to a week as the emergency services failed to make it through the rubble-filled streets to the worst hit areas.

There is no doubt that the Earth will move again for many hundreds of millions of people living in the world's most seismically active zones. In the past, individual quakes have killed thousands of people. Sadly, until the importance of suitable construction and effective preparation are taken seriously, many more will die.

**OPPOSITE Despite strictly enforced building codes, parts of San Francisco still sustained serious damage during the 1989 Loma Prieta quake.
BELOW Rescuers listen for survivors trapped beneath the rubble of their homes in the Indian state of Gujarat, following a devastating quake in January 2001.**

Europe's greatest loss

>>> On a sunny morning in late Autumn 1755, ancient Lisbon basked in the glory of its position as one of the world's great cities.

> Four days later, the elements of earth, fire, and water had conspired to wipe it from the face of the Earth. The heart of an empire and the home of the Portuguese fleet, Lisbon occupied a prime position on seven hills near the mouth of the Tagus river. In the middle of the eighteenth century, the city was a bustling, thriving port of close to 300,000 people, rich in culture and enjoying the fruits of its ever-growing colonial wealth. But shortly after 9:30 on the morning of All Saints' Day, the city was struck by one of the most destructive earthquakes to hit Europe.

As thousands bowed their heads in prayer in the central cathedral, a low rumble grew rapidly in strength. Within seconds the ground started to buck and shake, throwing worshipers on their faces and dislodging masonry from roofs and statues from their pedestals. With a great roar the full force of the quake struck with a speed that left no time for thought or escape. Like a pack of cards, the capital's great cathedral fell upon those below, crushing every living soul.

Total devastation

Many earthquakes are over in seconds, but the Lisbon quake just went on and on. Observers reported two separate quakes three minutes apart followed by a third about 12 minutes later. So intense was the ground-shaking over this period that virtually every building was reduced to a pile of rubble. Of a total building stock of around 20,000, it is estimated—perhaps conservatively—17,000 were completely destroyed. There were no seismographs at the time to measure accurately the scale of the quake, but it is now thought—from analysis of damage and contemporary observations—that it

may have registered as high as 8.75 on the Richter Scale. This would make it the largest known quake in Europe and one of the most powerful anywhere in the world in recent centuries. So violent were the shocks that they were felt across the whole of Portugal and much of Spain, and even as far away as North Africa.

The dazed and petrified survivors pulled themselves from the rubble and staggered into the debris-filled streets only to face another terrifying sight. Barely an hour after the main shocks, three enormous tsunami crashed into the harbor and the seafront, smashing ships to kindling and battering the waterfront buildings and any survivors sheltered within. Even as the waters began to recede a third peril made itself known as the first wisps of smoke started to drift across the city. Although the main public buildings were built from stone, most homes were of wooden construction. These and thousands of injured trapped within were

TOP AND ABOVE Over three-quarters of Lisbon's buildings were completely destroyed by the great offshore quake of 1755.

consumed greedily by raging firestorms ignited by countless smashed oil lamps, toppled candles, and overturned stoves. When the conflagrations eventually burned out, at least 30,000 of Lisbon's inhabitants were dead; some say as many 75,000.

Considering the event happened almost 250 years ago, the records of the impact of the quake are of quite an extraordinarily high standard. This is all due to one man, Sebastião José de Carvalho e Mello, the future Marquis de Pombal, and the minister responsible for rebuilding the city. Pombal arranged for a questionnaire to be sent to every parish in Portugal requesting detailed information on the quake and its effects, including the timing and duration of shocks. The catastrophe also triggered much speculation among the European scientific elite about why earthquakes occur. Dutch scientist J. F. Dryfhout was convinced that Lisbon had been leveled by the detonation of "explosive veins" deep within the Earth, but it was the Englishman John Mitchell who got it right, attributing earthquakes to the "shifting of masses of rock miles below the surface."

BELOW Tsunami and conflagration obliterated what little remained standing, contributing to a death toll that may have been as high as 75,000.

Firestorm

>>> When the founding fathers decided to establish a new city on the shores of a spectacular bay on the coast of northern California, they were quite unaware of the local geology and the great threat it posed to the city's future.

> Almost every school kid now knows that San Francisco sits astride the San Andreas Fault—one of the most active on the planet. Since dinosaurs roamed the Earth, the Pacific side of the fault has moved northward by 300 miles (500km) relative to the North American continent to the east, and this movement continues at 2 inches (5cm) a year—about the width of three fingers. Sometimes, however, parts of the fault lock for decades or even centuries, triggering violent quakes when they eventually let go. In 1906, although just 60 years old, San Francisco was already a great and proud city—a major commmercial and financial center.

Torched in minutes

In the early hours of April 18, 1906, the city's inhabitants were jolted awake by a huge magnitude 8.25 quake that in next to no time released a blast of energy greater than that generated by all the munitions used in World War II. Buildings great and small, new and old, tottered feebly and crashed to the ground, burying the streets and anyone on them at such an early hour beneath huge piles of rubble. Cable car rails were torn and twisted into fantastic shapes, while water and gas gushed from fractured mains. After three minutes the shaking stopped, but the real destruction of San Francisco was about to start. Despite the construction of many splendid mansions, particularly on famous Nob Hill, San Francisco was essentially a city of wood, and many of the older and poorer neighborhoods provided perfect fodder for the greedy flames that followed the quake. In all, perhaps 90 percent of the city's buildings were either built of wood or had wooden frames.

Overturned stoves and oil lamps, fractured gas mains, and live power cables all contributed

ABOVE The huge magnitude 8.25 San Francisco earthquake struck about 40 miles (65km) north of the city. BELOW Many buildings collapsed due to shoddy workmanship, but fire was the real destroyer.

THE CITY IN FLAMES
Photograph 30-31

A remarkable panorama of San Francisco in flames taken from Mason Street, 10:00 A.M. April 18th, 1906, 8 hours after the earthquake. No one looking at this picture would for a moment consider he was viewing a city "destroyed by earthquake." It requires very careful examination to locate the few evidences of earthquake damage. Even a very few damaged chimneys can be found. Only a few parapet walls have been thrown down. The elaborate ornamentations of Grace Church and Temple Emanu-el appear unharmed. It will be remembered that with the exception of a limited number of the taller buildings all of the masonry structures were carried on brick bearing walls. Note the many buildings such as the old St. Mary's Cathedral, Grace Church, Mills Building, Crocker Building, Palace Hotel, Mutual Bank, California Building, etc. which are entirely intact and apparently little harmed yet in the direct path of the flames which later partially destroyed them.

to over 50 fires that were raging within 30 minutes of the end of ground-shaking. Soon, strong winds had encouraged individual fires to merge into huge firestorms that ate their way across the city at a prodigious rate. Just one hour after the quake, the Postal Telegraph Cable Company got out the first message to the rest of the world:

"...earthquake at five fifteen this morning, wrecking several buildings and wrecking our offices. They are carting dead from the fallen buildings. Fire all over town. There is no water and we have lost our power..."

By the evening the downtown business district and Chinatown had been consumed by the flames, which were by then licking around the great mansions and hotels of Nob Hill. Fighting a losing battle against the conflagration were 585 firemen from 80 fire stations. Leaderless, because their chief had been killed in the quake, the men tried at first to contain the fires, but with their water supplies largely cut off by broken mains there was little they could do. The firestorms continued to rage throughout the following day, and it seemed at one point as if the whole city would go up in flames. Undoubtedly this would have happened if an army officer—General Frederick Funston—had not taken control of the firefighting operation.

In order to save the newest and most expensive properties in the city, the General ordered a stand to be made on Van Ness Avenue, one of San Francisco's widest thoroughfares. This he had widened to over 160 feet (50m) by using explosives to demolish houses—some of the finest in the city—along its length. The firebreak worked, and aided by a change in wind direction at least part of the city was saved. Nevertheless, when the smoke cleared on the morning of April 21, almost 4 square miles (10 sq.km) of "the American Paris" had been ravaged by the fires, with over 28,000 buildings destroyed.

ABOVE The rails of San Francisco's famous cable cars were torn and twisted by the severe ground-shaking.
BELOW The city hall collapsed along with many buildings in the business district, which consisted mainly of brick and stone masonry structures.

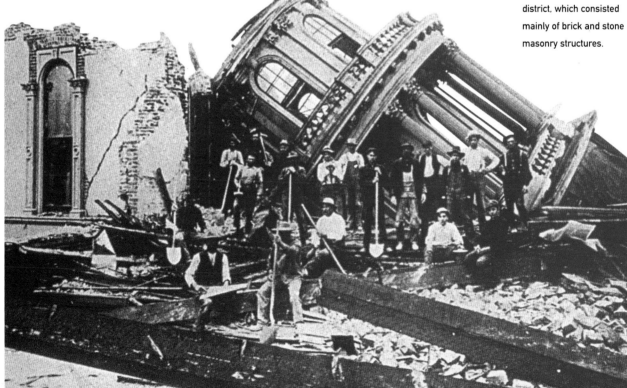

Turning point

>>>September 1, 1923 seemed just like any other early
Autumn day in the Japanese capital.

> The weather was fine and dry with just a hint of a breeze. As Tokyo's residents settled down to lunch, few suspected that it might be their last. Cafés and beer halls were brimming with hungry workers while in every home families prepared for the midday meal, oblivious to the imminent threat from below. At 11:58 a.m., a major fault beneath Sagami Bay, 50 miles (80km) to the south, ripped apart, generating an earthquake that registered 8.3 on the Richter Scale. Almost instantly the city of Yokohama was battered to the ground and just 40 seconds later the seismic shock waves crashed into Tokyo. Within seconds bustling cafés became flattened ruins, entombing the dead and the living. Many of those who, by apparent good fortune, did manage to fight their

way into the street, clung to life for only a few seconds longer before being crushed by debris plummeting from buildings high above. The shaking was so severe that people were hurled through windows and into the street, some to be swallowed up by fissures that snapped open, and closed as quickly.

As suddenly as it started the shaking stopped, and an eerie silence settled across the battered landscape, broken only by the periodic crash of debris from tottering and swaying buildings and the cries and screams of the trapped and

BELOW A highway torn apart by the tremendous forces unleashed by the 1923 Tokyo quake.

injured. Soon, however, another awful sound became apparent—the whispering and crackling of flames. While many thousands of deaths resulted from collapsing buildings during the quake itself, the real killer of the Great Kanto Earthquake was fire.

Home fires burning

In homes all across the city the ground-shaking had overturned countless *hibachis*, the open charcoal burners on which meals were prepared. In no time at all, tens of thousands of tiny fires had merged to form walls of fire that marched across the city incinerating everything in their paths. The screams of those incarcerated in the rubble combined with the hysterical wailing of crowds trapped helplessly between rapidly encroaching conflagrations. With nowhere else to go, thousands of dazed inhabitants headed for open spaces that they thought would provide refuge from the raging firestorms on every side. They were wrong.

On the banks of the Sumida river in an area known as Honjo, terrified survivors crammed themselves into an open space designated by the authorities as a refugee assembly area. By late afternoon the place was packed with a mass of humanity—over 40,000 people—completely surrounded by towering walls of flame. As the

burning of oxygen created a vacuum over the crowd, tornado-like winds were generated that sucked the fires together to form a gigantic firestorm that burned alive virtually every living thing in seconds. Only a few appallingly injured survivors staggered away. People were so tightly packed that many died on their feet, held upright by the bodies of those around them.

The firestorms continued to rage for nearly two days, leaving the once-great metropolis a shell of smoldering rubble. In total, some 360,000 buildings were destroyed in Tokyo and Yokohama, including 20,000 factories, 1,500 schools, and the Imperial University Library, one of the world's greatest depositories of old books and works of art. Some estimates put the final death toll as high as 200,000.

The devastation wrought by the Great Kanto Earthquake had more than a physical impact on Japan and its capital. The huge cost of the worst natural catastrophe in the country's history—around $50 billion in today's prices—was a terrible drain on the economy. A combination of the quake and the great stock market crash of 1929 plunged the country into the depths of economic collapse and depression. As in Weimar Germany between the wars, the climate of despair was hijacked by those who promised to make Japan a great and wealthy country again.

ABOVE LEFT So few buildings remained standing that survivors were forced to live and cook in the streets.
ABOVE RIGHT By the time the great firestorms had finally burned themselves out, 360,000 buildings were nothing but charred rubble.

Nature's debt

>>> It seems a sad fact of modern life that the great majority of major earthquake disasters afflict developing countries.

> This is not always the case, however, as the devastation caused by the Kobe earthquake of 1995 shows. Apart from a few early risers, most of the 1.5 million people of Kobe, Japan were deep in sleep when the ground started to tremble at 5:46 a.m. on January 17. The shaking was caused by a fault tearing apart just 9 miles (14km) beneath the surface and only 12 miles (20km) southwest of the city.

As the tear ripped northeastward it generated strong seismic waves that hammered into downtown Kobe within seconds. Sweet dreams were dramatically interrupted, causing drowsy inhabitants to stagger from their swaying beds and seek shelter wherever they could find it. In seconds the quake was in full swing with the violent shaking of the ground bringing down

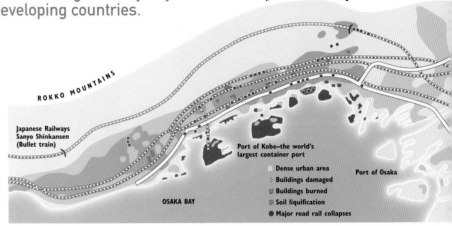

ROKKO MOUNTAINS

ROKKO MOUNTAINS

Japanese Railways
Sanyo Shinkansen
(Bullet train)

Port of Kobe—the world's
largest container port

Port of Osaka

Dense urban area
Buildings damaged
Buildings burned
Soil liquification
Major road rail collapses

OSAKA BAY

buildings all over the city, crushing many people before they were fully awake.

In the suburbs, the traditional Japanese buildings fared the worst as the severe shaking caused the weak outer walls and paper internal walls to cave in under the weight of heavy roofs made of tiles and mortar. With little warning, thousands of these buildings crashed to the ground all over the city, killing many elderly residents who slept on the ground floors.

Downtown, more modern office buildings fared better, but many were severely damaged. Debris falling from tall buildings made roads impassable. The elevated highways swayed and shook as the seismic waves continued to race through the city. While most survived intact, a section of the Hanshin Expressway succumbed to the battering, and the supporting columns shattered causing the road to collapse, killing 18 beneath its enormous mass. As the violent shaking continued, railway lines buckled, tunnels collapsed, and trains derailed. Fortunately, the quake occurred 14 minutes before the first of the famous 140-miles (230-km) per hour bullet trains took to the rails. Each bullet train holds

ABOVE Damage from the quake and the later fires stretched for a distance of over 90 miles (150km) along the north coast of Osaka Bay.
FAR LEFT Older buildings fared badly during the 1995 Kobe earthquake.

1,300 people and takes 2 miles (3km) to stop once moving at top speed.

Although the most severe shaking lasted a mere 17 seconds or so, this was quite sufficient to rupture gas and water mains and bring down power lines. In no time, fires were raging. Older houses, constructed largely of wood and paper, provided fuel for discarded cigarettes, candles, and overturned cooking stoves. By the time the fires were doused over 7,000 homes had been razed to the ground and many others severely damaged. Only the clement weather and lack of wind prevented a terrible inferno like the one that engulfed Tokyo in 1923.

No immunity

While the Great Hanshin Earthquake—as it is now known—was less apocalyptic than those that occurred in earlier times or in less prepared countries, it was nevertheless both lethal and devastating. Over 5,000 people died and a further 28,000 were injured. More than 140,000 buildings were destroyed or damaged and over 300,000 people were made homeless.

The quake teaches us an important lesson. Even major cities in industrialized countries are not immune from the effects of a devastating quake, especially those that occur close by and at a shallow depth. If the weather had been windier, both the death toll and the level of destruction could have been much greater; as it was, the performance of the emergency services was far from impressive. Fires burned out of control because fire engines could not get through the debris-blocked streets, while many survivors—some old or injured—had to fend for themselves for days before help arrived. At $200 billion, the economic cost of the quake made it the most costly natural catastrophe of all time. This, however, is likely to pale in significance when the next major earthquake strikes the Japanese capital.

LEFT AND BELOW Roads and railway lines suffered serious damage. Most notable was the collapse of part of the Hanshin Expressway, crushing 18 people to death.

Anatolia's fault

>>> The residents of northern Turkey have lived for thousands of years with the constant threat of a major earthquake, but somehow it still seems to take everyone by surprise.

> Slicing through the entire country, the North Anatolian Fault is one of the longest—at 900 miles (1,500km)—and most active fault systems in the world. Like California's San Andreas, it is a plate boundary, separating the giant Eurasian Plate from the much smaller Anatolian (Turkey) "microplate." As Africa and Arabia push northward toward Russia, Turkey is being squeezed sideways, causing it to scrape westward against the Eurasian Plate to the north at rates of up to an inch (3cm) a year.

Ever since the 1930s, the North Anatolian Fault has been rupturing in segments, each rupture generating a destructive quake, and each occurring further and further to the west. Big earthquakes, just a few hundred miles to the east in the 1950s and 1960s, must have played on the minds of the residents

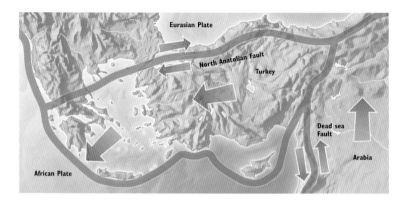

of Izmit for many years, particularly as the town had been destroyed at least seven times between 69 A.D. and 1719. On August 17, 1999, at 3:00 a.m., their turn came with a vengeance.

When the quake struck, nearly all of the 65 million residents of the Izmit and Istanbul provinces were asleep in their beds, resulting in an enormously high death toll. Due to a housing boom that had been going on for decades, many people lived in reinforced concrete apartment buildings up to seven stories high. Unfortunately the boom had attracted unscrupulous and corrupt construction companies that paid little regard to the government's earthquake building codes, with the result that the apartment complexes became death traps. In the severe ground-shaking that afflicted an area of over 770 square miles (2,000 sq.km), more than 6,000 buildings collapsed. Many experienced "pancake" failure, in which the walls fall outward and the floors crash down one upon another leaving little hope of survival for anyone inside.

In some areas, buildings simply sank down into soft sediment underneath as the seismic

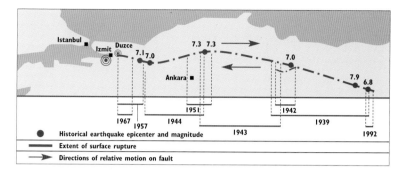

Istanbul ■
Izmit ■   Duzce
         ◉   7.1 7.0   7.3 7.3   7.0
        Ankara ■   7.9 6.8

                              1951   1942
        1967   1944   1939
             1957   1943   1992

● Historical earthquake epicenter and magnitude
━ Extent of surface rupture
→ Directions of relative motion on fault

OPPOSITE TOP The Turkey tectonic plate is being squeezed sideways like a grape in a vice.
OPPOSITE BELOW Rescuers reach a badly injured survivor of the 1999 Izmit earthquake. Over 17,000 others were less fortunate.

ABOVE Since the 1930s, destructive earthquakes caused by rupturing of the North Anatolian Fault have migrated further and further west—toward Istanbul.
BELOW One of over half a million people made homeless by the quake ponders an uncertain future.

shaking made the soil flow like water. A particularly poignant story recounts how a man who was unable to sleep woke his wife in the early hours and asked her to make some coffee. As they drank the beverage on their second floor balcony the quake struck and the building swayed and sank into the soft soil beneath. At one point it tilted so much that their balcony touched the ground and the couple was able to scramble out and run for safety. Then the apartment building pulled itself upright once more and collapsed like a deck of cards—killing everyone left inside.

Urban plight

The magnitude 7.4 Izmit earthquake was one of the few major quakes to strike a highly urbanized and industrialized part of the world in the last half century, and the scale of the disaster was truly terrible. The official death toll stands at around 17,000, but many more remain unaccounted for. Over 120,000 homes were destroyed or severely damaged and 500,000 people made homeless. Industry was also devastated. The Istanbul–Ankara motorway suffered heavy damage, communications cables were severed, and the country's largest oil refinery was reduced to an inferno for six days.

Meanwhile, the government is already looking nervously westward to the sprawling megacity of Istanbul, where 12 million people are crammed into dense estates of apartment buildings built no better than those of Izmit. No one knows exactly when Istanbul will be hit, and arguments rage about just how big the quake will be when it comes. What is certain is that the only part of the North Anatolian Fault left to rupture in the current earthquake cycle lies beneath the Sea of Marmara, south of Istanbul. When it finally goes, there seems little doubt that Istanbul will suffer the same terrible fate as Izmit.

SHAKE, RATTLE, AND ROLL

Rather like the town drunk, the Earth is constantly shaking. Fortunately for us, however, most of these tremblings are innocuous and release very little seismic energy.

Chile 1960

Alaska 1964

Aleutian Islands 1957

Kamchatka 1952

Kurile Islands 1963

All other earthquakes

Every year we experience around 100 serious earthquakes greater than magnitude 6 on the Richter Scale, capable of causing severe damage and great loss of life, particularly if they strike a poorly constructed and ill-prepared city. Really huge quakes are even less common, with only one or two magnitude 8 events every year, most of which occur far from areas of high population. Although rare, these huge earthquakes are important because they release almost all the seismic energy pent up in fault zones across the planet. Between 1904 and 1986, for example, the Earth was shaken by approximately 7 million quakes, but well over 50 percent of all the seismic energy released over the period was as the result of just 5 enormous earthquakes. Most notably, over a quarter of all the energy was released in a single seismic event —the Great Chile Earthquake of 1960.

One of the reasons why earthquakes are so destructive is that they are confined to the outer rigid shell of the planet—the lithosphere. If they occurred deep down within the Earth's interior we would hardly feel them, but there the rock is too hot to snap suddenly and instead deforms slowly rather like toffee. All other factors being equal, the most destructive quakes occur due to huge rock masses jerking spontaneously past one another along a shallow fault. These quakes occur in the top 12 miles (20km) of the crust and over 75 percent of the seismic energy released every year comes from earthquakes in

the outer 10 miles (15km) of the Earth.

Although we are no closer to predicting earthquakes, we are now much better at detecting them and a worldwide network of seismographs ensures that a major quake can be pinpointed and its size estimated within minutes of it happening. The worry is that, in the not too distant future, seismic stations all over the world will record a major, shallow quake directly beneath one of the world's great cities—maybe Istanbul, or Mexico City, or perhaps Tokyo. For the distant observers it will be hard to imagine that the trembling recording pen might be writing the death warrant of millions.

ABOVE Five earthquakes released over half of the world's seismic energy between 1904 and 1986. OPPOSITE Craftsmen build temporary shelters after the 1908 Messina earthquake in Sicily. BELOW Collapse of the heavy stone buildings of Messina led to the deaths of 120,0000 men, women, and children.

Measuring moving ground

>>> Interest in just what causes earthquakes is nothing new. Almost 2,000 years ago the ancient Chinese were already keen to try and understand just what made the ground beneath periodically shake and tremble.

> As early as the second century A.D. the ancient Chinese had developed and built an earthquake-recording device which, admittedly, bore little resemblance to a piece of scientific equipment. The apparatus consisted of a circle of eight bronze dragons aligned above a ring of eight bronze toads. Each dragon held in its mouth a small metal ball which, in the event of an earthquake, would be dislodged and fall into the gaping mouth of the toad sitting below. The theory would seem to have been that only the ball held by the dragon oriented in the direction of the quake would be dislodged and fall into the mouth of the corresponding toad, who would therefore point out the direction from which the quake originated. A nice idea—but a major quake would almost certainly have dislodged all the balls and probably projected the contraption through the nearest window.

The birth of the seismoscope

Across the other side of the world in western Europe there seems to have been little interest in the science of earthquakes until some 1,500 years later, when the seismoscope was invented. Like its earlier Chinese counterpart, this could hardly be described as a scientific gadget, consisting as it did of a tube of mercury that would spill out if the ground shook. A useful earthquake monitor did not, in fact, appear on the scene until the middle of the nineteenth century, when scientists realized that the pendulum could be used to amplify and record the often tiny reverberations coming from far-off earthquakes. These forerunners of the modern seismographs were simple devices consisting of a recording pen attached to a heavy pendulum suspended above a chart in such a way that they just touched. When an earthquake shook the ground the mass of the pendulum caused it to remain stationary while the Earth moved beneath it, causing the pen to trace out the path of the movement on the chart. Not surprisingly, such pendulum-based seismographs were developed and perfected in Japan and Italy— two countries prone to severe and destructive quakes. They allowed the ground-shaking during a quake to be recorded in three directions; north–south, east–west, and vertically. By the early twentieth century many such instruments were in use across the world, just in time to record the dramatic rumblings of the Great 1906 San Francisco Earthquake.

During the last 100 years the seismograph has become a much more sophisticated piece of

FAR LEFT The first earthquake-recording device, constructed by the ancient Chinese, bears little resemblance to modern seismographs. BELOW Ground-shaking can be detected and displayed on seismographs thousands of miles or kilometers from the quake itself.

hardware. Ground motions are now viewed on a monitor and analyzed by computers, which measure not only the components of ground movement, but other parameters of the seismic waves such as amplitude and wavelength, and provide much more information about the quake and its origin. An enormous global network of seismographs now exists, which can pinpoint the location of any quake in the world within minutes of it happening. This is vital where there might be a tsunami threat or where a potentially destructive quake has occurred in a poorly accessible area.

Although modern seismographs can tell us pretty much all we want to know about an earthquake, including its location, depth, strength, and the duration of shaking, they can tell us nothing about the damage incurred. Clearly, a large, shallow quake in an urban area is liable to

The Modified Mercalli Scale for Earthquake Intensity

SCALE	INTENSITY ON MODIFIED MERCALLI SCALE	EFFECT OR DAMAGE	CORRESPONDING MAGNITUDE ON RICHTER SCALE
I	INSTRUMENTAL	NOT FELT BY HUMANS	–
II	FEEBLE	FELT BY SOME PEOPLE AT REST	–
III	SLIGHT	HANGING OBJECTS SWING; SIMILAR TO HEAVY TRUCK PASSING	<4.2
IV	MODERATE	DOORS, WINDOWS, AND CROCKERY RATTLE; FELT BY PEOPLE WALKING	–
V	SLIGHTLY STRONG	DOORS SWING, LIQUIDS AND PICTURES DISTURBED; SLEEPERS WAKE; CHURCH BELLS RING	<4.8
VI	STRONG	WINDOWS AND CROCKERY BROKEN; WALKING DIFFICULT; TREES SWAY; MASONRY MAY CRACK	<5.4
VII	VERY STRONG	FURNITURE BROKEN, WALLS CRACK, AND PLASTER FALLS; NOTICED BY DRIVERS; DIFFICULT TO STAND	<6.1
VIII	DESTRUCTIVE	PARTIAL COLLAPSE OF POORLY CONSTRUCTED BUILDINGS; CHIMNEYS FALL; STEERING OF CARS DIFFICULT	–
IX	RUINOUS	SOME HOUSES COLLAPSE; UNDERGROUND PIPES BREAK; OBVIOUS GROUND CRACKING; GENERAL PANIC	<6.9
X	DISASTROUS	MANY BUIDINGS DESTROYED; LANDSLIDES AND SOIL LIQUIFACTION COMMON, MANY GROUND CRACKS	<7.3
XI	VERY DISASTROUS	MOST BUILDINGS AND BRIDGES COLLAPSE; RAILWAYS, PIPES, AND CABLES DESTROYED	<8.1
XII	CATASTROPHIC	TOTAL DESTRUCTION; TREES UPROOTED GROUND RISES AND FALLS IN WAVES; OBJECTS THROWN INTO THE AIR	>8.1

be both destructive and lethal, but much will depend on other factors, such as how well the buildings are constructed and how well the inhabitants and the emergency services are prepared. Using information from seismographic records, any quake can be allocated a position on the Richter Scale of earthquake magnitude, but this again will not provide categorical information on its impact. The Modified Mercalli Scale of earthquake intensity successfully fulfills this role by classifying a quake according to the damage it causes rather than its strength.

ABOVE The Modified Mercalli Scale classifies a quake according to the damage it causes, rather than its strength.

Waves of death

>>> It is now over 65 years since Charles Richter developed his scale for measuring earthquake magnitudes.

> The scale is logarithmic, meaning—in simple terms—that every increase in magnitude represents a tenfold rise in the amount of ground motion experienced. Thus a magnitude 7 earthquake is ten times more violent than a magnitude 6, while a magnitude 8 is 100 times more violent. Although the scale is open-ended nothing has yet been recorded that equals or exceeds 9, and hopefully it never will be. Richter's method is based upon the total amount of energy released during an earthquake, which can be determined from the maximum shaking as indicated by the pen trace on a seismograph record.

Measuring up

A full half-century before Mr. Richter announced his scale, the Italian seismologist Giuseppe Mercalli dreamed up a different way of determining the size of an earthquake. A modified version of Mercalli's original 1883 Scale (the Modified Mercalli Scale seen on page 107) is still very much in use. It is quite different from Richter's Scale, being based on direct observations of quake damage and other effects, rather than on the seismographic record. Mercalli's Scale does hold a number of advantages over Richter's, in that it does not require access to seismographic records and the scale of a quake can be determined some time after it has happened. Mercalli's method can even be used to estimate the sizes of quakes that have occurred centuries ago. But there are disadvantages too—the greatest being that Mercalli's Scale cannot distinguish between damage caused by a moderate earthquake close by and damage caused by a large quake further

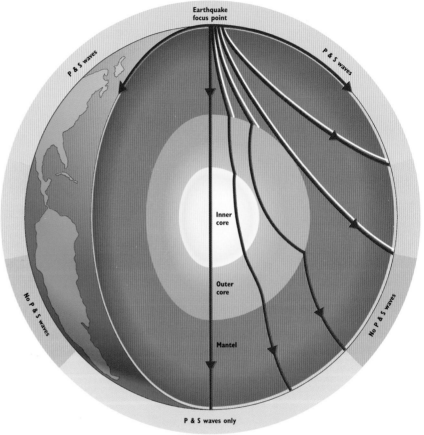

away. Together, however, the magnitude scale of Richter and Mercalli's intensity scale are able to provide an excellent picture of the energy released during a quake, the degree of ground-shaking, and the amount of damage caused.

The Richter and Modified Mercalli Scales measure, respectively, the degree of ground-shaking and its effects on buildings and other structures. The shaking itself is a reflection of the seismic waves that race outward in all directions when a fault suddenly jolts into life. Although they may appear as random squiggles to the

ABOVE Primary (P) and secondary (S) waves travel through the Earth's interior, arriving before the destructive surface waves that take the longer route around the planet's surface.

untrained eye, the traces made by the recorder pen on a seismograph chart can be clearly interpreted by earthquake scientists in terms of the different seismic waves. Some of these travel as shock waves through the atmosphere causing a characteristic rumbling or "express train" rushing sound, or—if the quake occurs beneath the sea—as potentially devastating tsunami. Most quake energy, however, is transmitted through the Earth and it is this the seismograph picks up.

When a fault moves it generates a huge amount of seismic energy that travels through the Earth as three different types of wave, referred to as P (primary), S (secondary), and surface waves. Each wave type travels at a different speed through the Earth and causes a different sort of ground-shaking, and therefore a different looking pen trace on the seismograph. Moving at 3.4 miles (5.5km) per second—about eight times faster than the Concorde—P waves, sometimes called "push waves," are the fastest of all, and therefore reach the seismograph first. They propagate by repeatedly squeezing and stretching the rock through which they move, in a manner similar to a line of wagons being shunted back and forth.

Hard on the heels of the P waves are the slower but far more damaging S waves, or "shake waves." These cause the rock they pass through to shake violently from side to side, turning roads and railways into waving ribbons of tarmac and steel. Both P- and S-waves travel through the body of the Earth. As their name suggests, however, surface waves take the longer route around the planet's surface. Consequently, they are the last to arrive, providing a real sting in the tail. Surface waves cause the ground surface to roll like the surface of the ocean, making bridges and skyscrapers sway back and forth and damaging gas pipes, power cables, and communication lines.

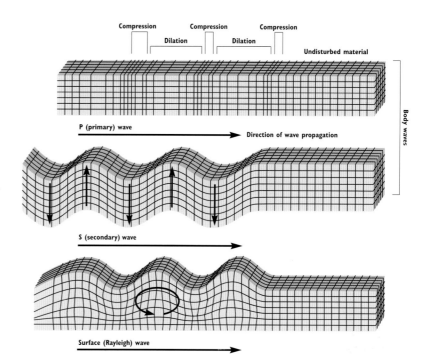

Compression   Compression   Compression
Dilation   Dilation
Undisturbed material
Body waves
P (primary) wave   Direction of wave propagation
S (secondary) wave
Surface (Rayleigh) wave

ABOVE P, S, and surface waves each cause the ground to shake in different ways.
LEFT The rolling motion of surface waves is particularly effective at damaging railway lines and roads, and at bringing down power cables.

Zeroing in

>>>For emergency relief to be mobilized rapidly, it is vital that the exact location of a potentially destructive earthquake is pinpointed as soon as possible, especially if the quake has struck an isolated part of the world.

BELOW If the distances to a quake epicenter can be determined from the records of three seismographic stations, then the position of the epicenter can be precisely pinpointed.

>A number of groups now exist that have the capability to do this, including the British Geological Survey's Global Seismology Unit, the U.S. Geological Survey's Global Seismic Network, and the French GEOSCOPE. But just how do they manage this?

The principle of underlying quake location is quite straightforward and is dependent on the fact that P and S waves travel at different speeds through the Earth. Consequently, the further away an earthquake is from a seismograph, the greater the time gap between the arrival of the P waves and S waves will be. If a quake occurs immediately below the seismograph then the P and S waves will arrive at almost the same time. If, on the other hand, the quake occurs in Japan and the seismograph is in Scotland, there will be a much greater gap between the arrival times of the P and S waves. This relationship means that just by looking at the seismographic record, the distance of any earthquake from a particular seismograph can be determined. But on its own this information is ambiguous and does not provide information about the direction of an earthquake relative to the seismograph.

If, however, there are three widely spaced seismographs—based in observatories in different countries or continents—then determining the distance to each will make it possible to locate the quake epicenter, the point on the surface directly above it. Pinpointing the precise position of the quake itself beneath the

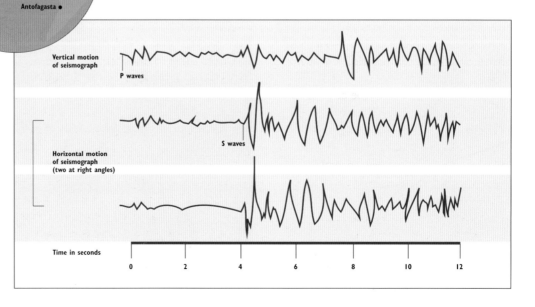

Golden, Colorado

Honolulu

Epicenter (Veracruz, Mexico)

Antofagasta

RIGHT The distance to a quake epicenter is determined from the time difference between the arrival of the P and S waves.

Vertical motion of seismograph

P waves

Horizontal motion of seismograph (two at right angles)

S waves

Time in seconds

0   2   4   6   8   10   12

surface—known as the focus—is a more convoluted process, but can be achieved with sufficient seismographs and computer power.

On-the-ground knowledge

Studying seismic waves a long way from their source can provide considerable information about an earthquake. The amount of energy released during the event can be determined together with the length of the quake. A strong quake that goes on for a long time might reasonably be expected to cause a lot of damage, but without knowing the local circumstances it is impossible to estimate from a remote location just how much damage. There are so many other factors besides strength and duration affecting how lethal and destructive an earthquake will be that on-the-ground knowledge is needed. Some of these factors are physical—for example, has a quake struck an

area of firm bedrock or one in which the geology comprises soft, weak sediments? Others are related to human activities. Is the building stock sound enough to withstand a quake? How prepared are the local population and the emergency services likely to be?

Due to such circumstances, any damage or casualty predictions based on remote records of earthquakes are likely to be wide of the mark. For example, while the magnitude 6.9 Loma Prieta earthquake that rocked San Francisco in 1989 killed only 64 people and caused a small amount of damage, the much smaller magnitude 5.9 quake that struck Agadir in Morocco in 1960 had a devastating impact, reducing much of the city to rubble and taking more than 14,000 lives. To get a real feel for the damage earthquakes can cause and why this can vary between quakes, we need to look at their effects and how these are influenced by both man and nature.

ABOVE Staff at the British Geological Survey's Global Seismology Unit in Edinburgh, Scotland, monitor earthquakes all over the world.

EARTHQUAKES AS DESTROYERS

During the last millennium around 8 million people were killed by earthquakes; some crushed and battered to death in collapsing buildings, the rest burned alive in fires, buried in landslides, or drowned by tsunami.

At the same time, the damage to social and economic infrastructure and destruction of property has cost quake-prone countries trillions of U.S. dollars at today's prices. A major quake can plunge an entire nation into the depths of despair, and—as happened after the 1923 Tokyo earthquake—can even alter its outlook. The impact on less powerful economies can be highly detrimental and last for many years. At $20 billion, losses due to the 1999 Turkish quake, for example, amounted to ten percent of the country's gross domestic product (GDP). The quake struck at a difficult time for Turkey, with soaring inflation and huge international debts. The massive cost of reconstruction and the reduction of GDP due to business disruption have made a bad situation considerably worse.

With more people on the planet and more megacities in high-risk seismic zones, there can be little doubt that the destructive effects of earthquakes will increase. A big problem lies in the fact that, while the number of poorly constructed dwellings built on marginal land around huge urban agglomerations increases, so the size of the earthquake needed to trigger a disaster goes steadily down. Tens of thousands have died in relatively small quakes over the last decade or so, as homes that are little more secure than a deck of cards succumb to even moderate ground-shaking. In 1960, a quake registering a mere 5.9 on the Richter Scale took over 14,000 lives in the Moroccan city Agadir. It

is inevitable that, in the near future, such modest quakes will begin to take a greater toll on both people and property, causing a rise in seismic disasters and increasing the misery in the shantytowns that are now appearing around quake-prone cities across the planet.

OPPOSITE The huge quake that struck the Indian state of Gujarat in January 2001 completely destroyed 400,000 buildings and may have killed as many as 100,000.

The most destructive earthquakes of the last 100 years

YEAR	COUNTRY AND REGION/CITY	RICHTER MAGNITUDE	FATALITIES	ECONOMIC LOSSES IN MILLIONS US$
1906	U.S. (CALIFORNIA, SAN FRANCISCO)	7.8	3,000	524
1908	ITALY (MESSINA)	7.5	25,926	116
1915	ITALY (AVEZZANO)	7.5	32,610	25
1920	CHINA (GANZU)	8.6	235,000	25
1923	JAPAN (TOKYO-YOKOHAMA)	8.3	142,800	2,800
1927	CHINA (XINING)	8.3	200,000	??
1936	PAKISTAN (QUELTA)	7.5	35,000	25
1939	CHILE (CONCEPTION)	8.3	28,000	100
1939	TURKEY (ERZINCAN)	8.0	36,740	20
1960	MOROCCO (AGADIR)	5.9	12,000	120
1970	PERU (CHIMBOTE)	7.7	67,000	550
1976	GUATAMALA (GUATAMALA CITY)	7.5	22,084	1,100
1976	CHINA (TANGSHAN)	8.0	290,000	5,600
1985	MEXICO (MEXICO CITY)	8.1	10,000	4,000
1988	ARMENIA (SPITAK)	6.9	25,000	14,000
1989	U.S. (CALIFORNIA, SAN FRANCISCO)	7.0	68	6,000
1990	IRAN (WESTERN)	7.7	UP TO 50,000	??
1994	U.S. (CALIFORNIA, LOS ANGELES)	6.7	61	44,000
1995	JAPAN (KOBE)	7.2	6,348	200,000
1999	TURKEY (KOCAELI)	7.4	19,118	20,000
2001	INDIA	7.7	UP TO 100,000	1,800

We all fall down

>>>When the ground is shaking so violently that it is impossible to stand or walk it is hardly surprising that many buildings soon totter and ultimately crash to the ground.

> Broadly speaking, the degree of shaking is greatest if the quake focus is shallow and close by. Thus, while the 9 mile (14km) deep Kobe quake, just over 12 miles (20km) from the city center, killed over 5000 and left 500,000 homeless, the similar size Loma Prieta quake that struck in 1989, over 60 miles (100km) from San Francisco, resulted in very few deaths and just 10,000 homeless. The duration of shaking is also critical, and the damage incurred at Kobe would have been considerably greater had the quake lasted longer than a mere 20 seconds.

Although destructive firestorms, tsunami, and landslides are triggered by severe earthquakes, the lasting image of a quake aftermath is one of the emergency services struggling to free those trapped in the mountains of rubble that once were office buildings, schools, and hospitals. A striking legacy of many large quakes, however, is the obliteration of some structures while others close by remain unscathed. The local geology is a critical factor in building survivability and those constructed on solid bedrock generally fare much better than others built on soft sediments, landfill, or reclaimed land.

This is because such unconsolidated material experiences so-called liquefaction when shaken by seismic waves, particularly if it is saturated with water. As a result, the ground literally behaves like a liquid and can therefore no longer support the weight of the buildings above, which sink and topple into it. During the catastrophic Armenian earthquake of 1988, seismic waves were amplified as they entered soft sediments, causing the ground to shake eight times more powerfully, bringing down dozens of modern apartment buildings and killing over 25,000 people. Liquefaction was also responsible, during the Loma Prieta quake, for collapse of part of the elevated Nimitz Freeway in Oakland and severe damage to buildings in the Marina district of San Francisco.

Minimizing the damage

Basically, there are only two ways of reducing casualties and damage arising from collapsing buildings during an earthquake: either abandon quake-prone regions or construct quake-proof buildings. With over 40 countries affected by earthquakes, the former is clearly not feasible, but much can be done to improve the survival rate of buildings. In developing countries many casualties result from the collapse of adobe (mud brick) buildings that quickly succumb to even modest ground-shaking. There is no way that the inhabitants of poor countries can afford to construct high quality, quake-proof, reinforced concrete buildings, but low-cost methods are now being tried that can make their dwellings less like death traps. These include avoiding heavy block walls and massive lintels above doors and windows; replacing them with light, timber-framed, cane walls covered with plaster and clay.

In developed countries such as the United States and Japan, even the severe ground-shaking caused by big earthquakes has little effect on buildings put up in the last few decades, reflecting the stringent enforcement of strict seismic building codes. A spectacular example is the San Francisco Marriott hotel, which opened on the very day of the Loma Prieta quake. Even though every glass in the building was

RIGHT The Marriott Hotel
in downtown San
Francisco was opened
on the day of the Loma
Prieta quake in 1989, but
survived the onslaught.
BELOW Buildings, not
earthquakes, kill people.

smashed—except one that they proudly display in a glass case—the structure itself suffered little damage and survived relatively unscathed. Modern buildings like the Marriott can be quake-proofed in a number of ways, including bracing the walls, providing the foundations with rubber shock absorbers, or resting a massive counterweight on the roof.

The problem remains, however, that many modern cities in the developed world have a range of building stock of widely different ages. Older properties in many Japanese towns and cities are often not up to modern standards, and so-called retrofitting—ensuring such buildings meet today's seismic code specifications—is an expensive business. In developing countries a major problem lies in the absence of seismic construction codes or—if they do exist—the lack of enforcement. Vulnerability remains very high and there is a long way to go here before earthquake casualty numbers start to decrease.

Moving mountains

>>> When a large earthquake strikes a region of hilly or mountainous terrain, the end result is often the triggering of devastating landslides that can be at least as lethal as the wholesale collapse of buildings.

> As the ground is shaken violently, so huge chunks of rock are dislodged from unstable slopes to crash down on unsuspecting settlements below. In 1970, a magnitude 7.7 quake off the coast shook apart the overhanging rock face of the Nevados Huascarán mountain in the Peruvian Andes, bringing down millions of tons of rock, snow, and ice onto the valley towns clustered around the foot of the towering peak.

Driven by gravity, such giant landslides falling from great heights attain enormous velocities, and within four minutes a wall of debris the height of a 25-story building and traveling at the speed of a jet aircraft crashed into the towns of Yungay and Ranrahirca and neighboring settlements. When the dust had settled, the towns and villages had completely vanished along with more than 25,000 of their inhabitants.

BELOW Workmen dig for survivors buried by the huge landslide triggered during the 2001 El Salvador earthquake.

Landslides are very much the poor relations of the tectonic hazard family and in any lists of death and destruction due to natural catastrophes their involvement is usually grossly underestimated. This is because landslides are primarily secondary hazards. In other words, they are triggered by another catastrophic event such as an earthquake, a volcanic eruption, or a period of torrential rainfall. Consequently, despite causing mayhem across the planet year after year, the true impact of landslides in terms of loss of life and the destruction of property is lost—hidden away in the doom-laden statistics of quake, eruption, and storm.

Lethal landslides

Earthquakes are particularly effective at generating lethal landslides, and in fact over half of all Japanese quake-related deaths over the last 35 years were due to landslides rather than collapsing buildings, fire, or tsunami. Going back further in history, one of the greatest natural disasters of all time resulted from landslides triggered by an earthquake.

Half a millennium ago, in the Shensi province of China, the vast majority of the population lived not in houses but in cave systems excavated within the local hills. The caves were particularly easy to dig because the rock was of a very soft and fine variety known as loess—the result of rock crushed and ground by glaciers to a flourlike consistency and then transported by the wind to its current location. Although ideal for homemaking, the loess hills became death traps during an earthquake. In 1556 the region was shaken by a powerful quake that caused the weak sediments making up the hills to flow like water. Within seconds, thousands of cave homes were buried deep beneath landslides and the lives of 800,000 people snuffed out.

Few lessons, it seems, have been learned over the past 500 years, and landslides continue to

devastate communities whenever earthquakes strike areas of steep terrain. On January 21, 2001, the central American country of El Salvador was hit by a 7.6 quake that destroyed 20,000 buildings and left 1.3 million people homeless. In Las Colinas, a middle-class suburb of the capital, hundreds of children and their parents were in the local playground when the quake hit. Within seconds, the violent shaking dislodged the hillside above, bringing down a gigantic mass of mud, rock, and over 400 houses that had been constructed on the steep hillside. Minutes later, the playground was a silent wasteland of mud scattered with children's toys but with no signs of life. Many had opposed the building of the new villas on the hillside, but the construction company had challenged the government in court, and won. As the need for living space increases so many more buildings will begin to go up on the steeper, marginal land that surrounds many cities. Without improved controls on where to build safely, it is inevitable that many more playgrounds and their occupants will go the same way as that in Las Colinas.

ABOVE The Las Colinas landslide in El Salvador destroyed 400 homes and obliterated a packed children's playground.

Return of the sea

>>> Waking up in the tropical paradise of New Guinea brings few surprises.

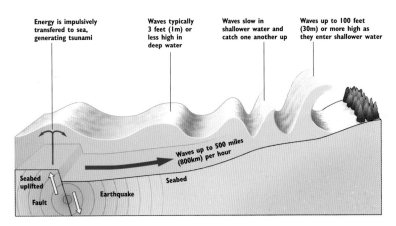

Energy is impulsively transfered to sea, generating tsunami

Waves typically 3 feet (1m) or less high in deep water

Waves slow in shallower water and catch one another up

Waves up to 100 feet (30m) or more high as they enter shallower water

Waves up to 500 miles (800km) per hour

Seabed uplifted

Seabed

Earthquake

Fault

> To the waking inhabitants of the villages of Sissano, Arop, Malol, and Warapu, shoehorned onto a narrow strip of sand between the Bismarck Sea and a palm-fringed lagoon, July 25, 1998 appeared to be a day like any other. For many, however, that dawn would be their last.

In the early evening, as villagers were settling down to a *sing-sing* to celebrate Papua New Guinea's independence day, the ground shook violently and cracks burst open and ripped along the length of the shore. The shaking stopped as quickly as it had started and the curious strolled down to the beach to examine the new cracks —only to be greeted by the terrifying sight of a huge wave, the height of a house, bearing down on them. The petrified onlookers had no time to run; they were picked up by the wave and carried back to their villages, where they were battered to pieces against their own homes. Two even bigger waves followed—as high as a five story building—tearing the palm trees from their roots and scouring the narrow strip of land clean of all signs of habitation. When the waters retreated, the villages were gone and the sands strewn with the dead and injured. No one will ever know how many were killed, but out of a population of some 10,000, nearly 3,000—many women and children—would never wake again.

Submarine landslide

The devastating tsunami that obliterated Sissano and neighboring villages on the northern coast of Papua New Guinea were generated by a moderate offshore earthquake. Such a quake would not normally be large enough to produce destructive tsunami. On this occasion, however, the quake caused the seabed to drop instantaneously by 6 feet (2m), causing a

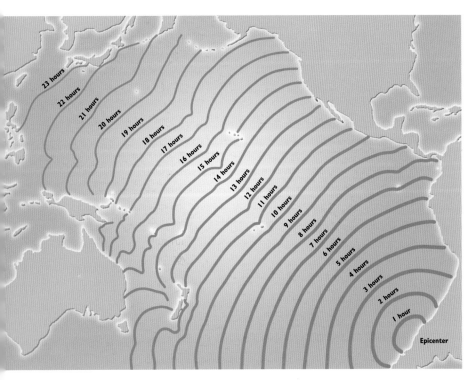

23 hours
22 hours
21 hours
20 hours
19 hours
18 hours
17 hours
16 hours
15 hours
14 hours
13 hours
12 hours
11 hours
10 hours
9 hours
8 hours
7 hours
6 hours
5 hours
4 hours
3 hours
2 hours
1 hour

Epicenter

TOP Tsunami are formed by submarine earthquakes, which impart a sharp jolt to the seabed.

ABOVE Waves from the massive 1960 Chile quake brought destruction to many parts of the Pacific over the following 20 hours.

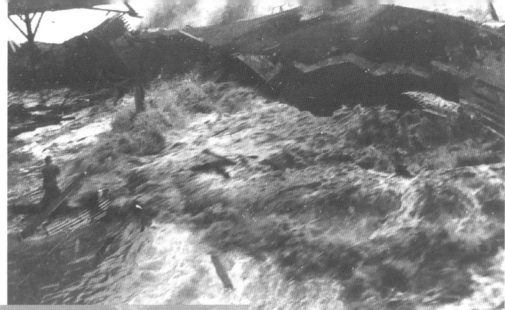

RIGHT In 1946, a tsunami crashes into the town of Hilo in Hawaii. No one who gets this close to a tsunami has any chance of survival.

Earthquake Magnitude compared to Tsunami Magnitude

EARTHQUAKE MAGNITUDE	TSUNAMI MAGNITUDE	MAXIMUM RUN-UP (METERS)
6	-2	<0.3
6.5	-1	0.5–0.75
7	0	1–1.5
7.5	1	2–3
8	2	4–6
8.25	3	8–12
8.5	4	16–24
8.75	5	>32

submarine landslide that in turn triggered the giant waves. Due to the enormous velocities of tsunami—which can move as fast as a jet aircraft—and because the quake occurred just off the coast, there was simply no time for the local population to seek cover. Even before villagers could react to the preceding quake, the waves were piling up along the shoreline.

In the original Japanese, tsunami means "harbor wave," a translation that sheds light on the behavior of these enormously destructive phenomena. In deep water, tsunami are barely detectable; their passage beneath the keels of a fishing fleet being marked by little more than a swell 3 feet (1m) high. As they enter shallow, coastal waters, the leading waves slow, causing the mass of water behind to pile up. In this way, an innocuous swell in deep water can become a towering wall of water 100 feet (30m) high as it approaches a harbor. Because they can travel at speeds of up to 500 miles (800km) an hour, the destructive range of tsunami is enormous.

Waves generated by the great 1960 Chile quake took only 15 hours to cover the 6,000 miles (10,000km) to Hawaii, where they destroyed 500 houses and drowned 60 residents. A full 20 hours after the quake, the tsunami battered the coasts of Japan—10,000 miles (16,000km) away—still with sufficient energy to take a further 180 lives. Unlike wind-driven waves, which have wavelengths of just a few dozen yards, the distance from the crest of one tsunami wave to the next is typically hundreds of miles. This means that when a tsunami hits it just keeps on coming for perhaps 15 minutes or so, carrying everything with it, before taking just as long to retreat. This raises their destructive potential enormously as houses, possessions, and people are smashed together in a maelstrom of debris, the living, and the dead.

Fire following

>>>The 1906 San Francisco and 1923 Tokyo quakes teach that when the Earth stops shaking the real problems often start.

BELOW The 7,000 buildings destroyed by fire after the 1995 Kobe earthquake bear testimony to the continuing threat of conflagration after a major quake strikes an urban center.

OPPOSITE LEFT Following the 1989 Loma Prieta earthquake, San Francisco once again became home to burning buildings, if on a smaller scale than in 1906.

>The conflagrations that often follow a major earthquake can be far more destructive than the quake itself, particularly if a city's buildings are primarily wooden or contain a high proportion of wood. As recently as 1995, when a large quake devastated the Japanese city of Kobe, fires were a major cause of destruction. Although the 7,000 buildings destroyed by fire in Kobe pale in significance when compared with the 28,000 destroyed at San Francisco, or the hundreds of thousands razed to the ground in Tokyo, they do bear testament to the continued vulnerability of modern cities to the post-quake fire hazard.

Whether or not a major fire problem develops after an earthquake depends upon many factors, some of which are perhaps more obvious than others. Most important, there needs to be available fuel and a means of ignition, both of which can be found in abundance amid the rubble and ruins generated by severe ground-shaking. Wood is an obvious fuel candidate, and numerous towns and cities still maintain a large stock of wooden or at least wood-framed buildings. Many other combustible and flammable materials, however, also contribute to quake-generated fires, including furnishings, plastics, chemicals, natural gas, and gasoline and related fuels. Sources of ignition also abound in severely damaged urban centers, including live power cables, candles and oil lamps, cooking stoves, and open fires. Fuel and ignition sources in close proximity will start small fires that will swiftly grow and merge into larger conflagrations provided a source of fuel is maintained and other conditions are favorable.

The last supper

One of these conditions is the weather. The spread of fires on a wet and windless day will obviously be hindered, while on a warm day with strong winds following a long dry period, major fires—once they have taken hold—will be almost impossible to control. To a large extent, the fire hazard after an earthquake depends upon the number of initial fires, which can in turn have close links with the timing of the quake. Particularly in developing countries, an earthquake that strikes during lunch or dinner is likely to start more fires by overturning cooking stoves and grills and scattering hot coals and burning wood. Similarly a quake during the late evening can contribute to more blazes by knocking over countless oil lamps and candles.

With an estimated 1 million wooden buildings still standing in Tokyo, the Japanese government is understandably concerned about the threat of fire when the next "big one" strikes. The last thing they want is a repeat of the appalling carnage caused by fires following the Great Kanto Earthquake in 1923.

Modern Japanese families now use far fewer stoves and grills than during the 1920s, but evidence from Kobe showed that fires can still start easily and spread rapidly. In a future Tokyo quake, gas leaks in homes and fractured gas mains have been identified as posing major fire threats and trying to limit these is currently a top priority. As part of this plan, the Tokyo Gas Company has developed what it calls an "intelligent" gas meter. Using its own tiny computer, the Miconmeter is able to shut off the gas supply if it detects a quake of intensity 5 on the Japanese Seismic Scale. The meter was introduced in the early 1990s and operated successfully when the city was struck by a magnitude 5.9 earthquake in 1992. The Miconmeter does not, however, address the problem of fractured or leaking gas pipes or the many other potential fire triggers, and it seems likely that when Tokyo next succumbs to a really big quake, the columns of smoke and flame will once again rise over the ruins of this great city.

BELOW Firefighters will struggle to control the great firestorms that can be expected when the next great quake strikes the Tokyo region of Japan.

SPOTTING THE BIG ONE

Just twice in the last 30 years have highly destructive earthquakes struck major urban centers in the developed world: In 1995 at Kobe, Japan, and four years later at Izmit in Turkey.

In both cases the devastation was immense, resulting in combined totals of well over 20,000 lives lost, 300,000 buildings destroyed or severely damaged, and 800,000 people made homeless. Notwithstanding these staggering statistics, with respective magnitudes of 7.2 and 7.4, neither Kobe nor Izmit can be regarded as major earthquakes. In light of their appalling consequences, however, concern is growing about the likely impact of a quake ten times as large on a highly urbanized and industrialized region in a developed country. Understandably, much attention is focused on Tokyo, razed to the ground by the magnitude 8.3 Great Kanto Earthquake in 1923, where a major quake is expected to strike in the next few decades.

But can we say when and where the next "big one" will strike? The answer is a categorical "no." We can pinpoint active faults whose quake histories are well documented and which are now due another quake, but we cannot say exactly when. Despite considerable research over the past half century, there has still not been a single case of successful earthquake prediction. Furthermore, there are other parts of the world where all the active faults have not yet been identified or where the historical earthquake record is poor. For such regions, seismologists have little idea—even in general terms—of when the next quake will occur or how big it will be.

Of particular concern are those parts of the planet that have experienced gigantic earthquakes in the past, but have been ominously quiet for the past few centuries. Two regions that immediately spring to mind are the New Madrid area of the central U.S. and the Seattle area of the U.S. Northwest, where huge quakes tore the crust apart in the eighteenth and nineteenth centuries. A major quake in a region that has not experienced one within living memory is especially worrying, as the affected communities are likely to be totally unprepared for the shock.

OPPOSITE Where next will the crust be torn asunder in the developed world? BELOW Geologists measure the amount of earthquake rupture after the 1995 Kobe earthquake.

Educated guesswork

>>>The terms forecasting and prediction are often used synonymously when applied to earthquakes and other natural disasters, although their meanings are in fact quite different.

>To anyone who has carried an umbrella around for a week of blazing sunshine, only to get soaked to the skin the next day having left it at home—all because of the assurances of a weather forecaster—it should come as no surprise that the term "forecast" is used to describe a relatively imprecise statement. "Prediction," however, refers to something altogether more precise and accurate, and covers a shorter time period than a forecast. Accepting this terminology, we can make quake forecasts—of a sort—although true prediction remains, for the time being, out of our grasp.

Forecasting earthquakes relies upon knowledge of the past record of quakes on a particular fault, or in a particular region, to identify a regular pattern that can provide a clue to the size, location, and timing of the next earthquake. This would be easy if quakes were regular. For example, if a quake on a certain fault had occurred in 1802, 1852, 1902, and 1952, then the inhabitants of settlements close by should be getting worried. Because nature does not operate like a ticking clock, however, a more likely sequence of quakes would be 1802, 1841, 1909, and 1952. The average return period in both cases is 50 years, but in the second series the time between quakes varies from 39 years to 68 years, making another quake certainly due, but perhaps not until 2020 or even beyond.

Where the historical earthquake record only goes back a few centuries, forecasts can be virtually meaningless. If, for example, the previous quake in the sequence was in 1630, then local inhabitants could have well over a century to wait before the Earth shakes again. The general

RIGHT The surface expression of the San Andreas Fault is clearly visible as it slices across the Californian desert. BELOW FAR LEFT A fault separates dark shales from pale limestone. Geologists can learn a great deal about a fault's past movement by studying such outcrops. BELOW The motion on the San Andreas Fault during the 1989 Loma Prieta earthquake.

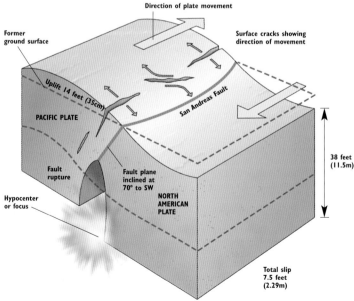

Direction of plate movement

Former ground surface

Surface cracks showing direction of movement

Uplift 14 feet (35cm)

PACIFIC PLATE

San Andreas Fault

Fault rupture

Fault plane inclined at 70° to SW

NORTH AMERICAN PLATE

Hypocenter or focus

38 feet (11.5m)

Total slip 7.5 feet (2.29m)

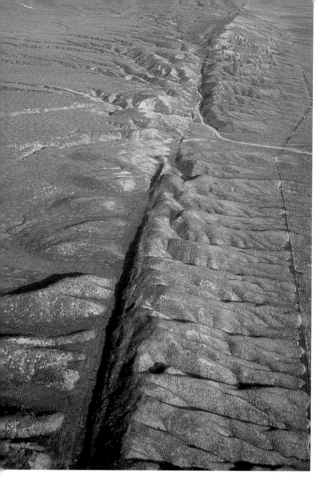

rule then is—the longer the record the better the forecast.

The future lies in the past

Most active fault systems are characterized by earthquakes that repeat over periods that are at least in the same "ballpark." This is because the sudden fault movement that triggers a quake occurs in order to release strain that has accumulated in the rocks. If strain is increasing at a constant rate and there is some threshold, determined by the material properties of the rock and the nature of the fault, beyond which the strain is relieved by fault movement, then some periodicity in the earthquake record can be expected. On many faults, strain accumulation is related directly to the movement rates of the

lithospheric plates, and faults like the San Andreas and Turkey's North Anatolian Fault actually mark the contacts between plates that are moving past one another at a constant few inches a year. If the annual rate of movement is known, along with the amount of fault movement that occurs during an earthquake, then the frequency of quakes can be deduced. If the lithosphere on either side of a fault is moving at an inch (2cm) a year, for example, and the fault jumps 6 feet (2m) during every quake, then the strain building up along the fault must be released every 100 years. The fault is actually trying to move at the same rate as the lithosphere but is prevented from doing so by friction. Every century, however, the strain becomes so great that friction is overcome, the fault jumps, and an earthquake is triggered.

Unfortunately the situation is not always this simple. Faults often occur not individually, but as part of complex systems. Sometimes when a fault ruptures to generate an earthquake, its movement actually stresses another fault in the system. This may push the strain in the neighboring rocks to such a level that that fault moves, causing a second quake. This stress transfer mechanism can result in several quakes occurring one after another over years or decades, and accounts for the steady and ominous east–west progression of quakes along the North Anatolian Fault—straight toward Istanbul.

RIGHT The positions of active faults and past earthquakes are used to compile a seismic hazard zonation map for New Zealand. The northern South Island and southern North Island are the most at risk, with damaging quakes expected every two decades or less.

LARGE EARTHQUAKES SINCE 1840
Magnitude (Richter Scale)
● 6.5–6.9
○ 7 and greater

Alpine fault

SEISMIC HAZARD REGIONS
Return period for a damaging earthquake
Less than 20 years
20–50 years
50–200 years
Over 200 years
/ Main fault lines

The Holy Grail

>>> There is no question that the accurate and repeatable prediction of earthquakes remains the Holy Grail of geophysics, guaranteeing a Nobel Prize to the successful scientist. But will it ever be possible?

> Three decades ago, quake prediction was very much in vogue, and the U.S. Geological Survey in particular devoted much time and money to monitoring the San Andreas Fault system in the hope of spotting clues to predicting future earthquakes. An armory of high-tech instrumentation was brought to bear to measure strain accumulation in the rock and to look for unusual movements that might indicate that a quake was on its way. Strain meters revealed pressures building up in the rocks around the fault, while creep meters and laser beams bounced off reflectors on the far side of the fault measured displacements as small as 0.04 inch (1mm). Scientists also dug wells and watched the water levels within. If cracking of the rock deep down preceded a big earthquake, then the water should enter these cracks causing the level to fall. At the same time, the naturally occurring radioactive gas, radon—which is common in groundwater—should be able to rise up through the cracks, so special radon detectors were also built to look for increases in the level of the gas.

While some focused on fault monitoring, others looked to their earthquake records to search for clues to coming earthquakes. Sometimes a major quake is preceded by a number of smaller foreshocks or even tiny quakes called microseisms. The problem is that neither always occur before a big quake, so they can't be relied upon. Even more confusingly they can occur without any succeeding major quake.

The optimism of the 1970s has now largely dissipated and many earthquake scientists in the United States and Europe feel that useful prediction remains as far off as ever. In Japan and China, however, reports of unusual animal behavior before a big earthquake—including excitable catfish and aggressive pigs—are taken seriously enough for scientists to look for possible causes. One suggestion is that electrostatic charges generated by accumulating strain in the rocks cause our furry and feathered

BELOW In addition to spectacular surface fractures formed during earthquakes, cracking deep down before a quake may lead to changes in well water levels and releases of radon gas.

friends to experience small but irritating electric shocks. It seems unlikely, however, that monitoring animal behavior will ever constitute a serious earthquake prediction tool. The method is simply too subjective and depends as much on what an observer defines as unusual as it does on the behavior of the animals themselves.

On a more serious level, huge controversy has surrounded a group of Greek seismologists who claim to be able to predict earthquakes by detecting beforehand what they call seismic electric signals. The three scientists, Varotsos, Alexopoulos, and Nomicas (collectively known as VAN) believe that the signals are caused by stress changes in the crust and can be detected weeks before an earthquake and hundreds of miles away. The VAN team claims to have predicted earthquakes in Greece, but its method remains highly controversial and has been described by seismologists as "absurdly vague."

Should we know?

It may seem like an odd question, but do we really want to be able to predict earthquakes anyway? Imagine for a minute what would happen if this were possible and the U.S. Geological Survey forecast in 2020 that San Francisco would be struck by a magnitude 8 quake two years later. The impact on the local economy would be devastating. Property prices would fall through the floor as potential buyers vanished, while insurance companies would be the next to go, leaving homeowners with no coverage. Companies of all sizes would desert, off-loading their workforces and leading to unprecedented unemployment levels and social upheaval. It may even be that the damage to the local economy could be greater than that likely caused by a quake itself. In these circumstances, perhaps it would be best if the Holy Grail of successful earthquake prediction were to remain just beyond our reach.

TOP Can the use of high-tech instrumentation and increasing computer power ever enable us to successfully predict earthquakes?

ABOVE If we could accurately predict earthquakes, the consequent devastating impact on the local economy could be as great as the quake itself.

Teaching and learning

>>> While true earthquake prediction may not become a reality for decades, if ever, there remains much we can do to limit the impact of large quakes when they strike in addition to constructing safer buildings.

> It is absolutely critical that communities at risk from destructive earthquakes are adequately prepared beforehand. This means an effective education scheme designed to ensure that people behave appropriately before, during, and after an earthquake. Getting the message across is not as easy as might be expected. Even in places like California, where the population is constantly bombarded with information about what to do when the Earth shakes, in the terror of the moment many forget what they have been taught. After the 1989 Loma Prieta quake that hit San Francisco, nearly 70 percent of people questioned said that they either froze or ran outside, with only 13 percent stating that they sought immediate protection.

China's Great Tangshan Earthquake killed 150,000 people in 1976, with the survivors being those who sought shelter under heavy items of furniture at the first hint of ground shaking. This appears to be the best advice that can be given to anyone caught up in a large earthquake, although other safe havens include doorway lintels and the corners of rooms, where even if roof collapse occurs, sufficient space may be preserved to allow survival. Other good advice given out as part of earthquake community preparedness programs encourages people to keep at hand an emergency pack, containing items such as water purification tablets, basic medication, matches, and a flashlight.

To minimize casualties it is also crucial that the emergency services learn to operate effectively. The moments immediately following a large, destructive quake are invariably filled

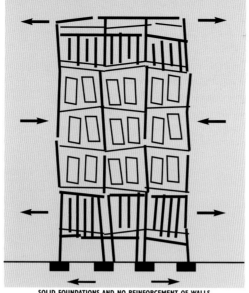

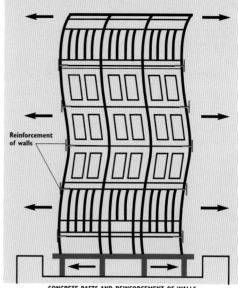

SOLID FOUNDATIONS AND NO REINFORCEMENT OF WALLS

CONCRETE RAFTS AND REINFORCEMENT OF WALLS

Reinforcement of walls

RIGHT New reinforced buildings constructed on movable foundations fare considerably better than unreinforced buildings on solid foundations. OPPOSITE Every year, on September 1—the anniversary of the Great Kanto quake of 1923—millions of school children and others across Japan take part in earthquake drills.

with chaos, and clear decision-making within the first few hours can save many lives. The length of time available for rescue teams to extract survivors from collapsed buildings is critical, as after 48 hours or so few are likely to be left alive. Many developing countries still place considerable reliance on special search teams—such as International Rescue—from developed countries. However, as these may take two days or more to arrive on the scene, it is vital that local teams are better trained in extracting survivors from often precarious piles of debris.

Planning for a disaster

If deaths and injuries are to be kept as low as possible, an effective disaster plan must be in place before the quake, and all the major players must fully appreciate their responsibilities and be familiar with their lines of command and communication. Without such pre-planning the aftermath of a quake will quickly deteriorate into confusion, disorganization, and chaos. The role of the emergency services is critical and they must be able to get their equipment rapidly to where it is needed. This clearly did not happen after the Kobe quake in 1995, when vital emergency supplies could not be moved due to roads being

blocked by debris. To prevent this from occurring, essential supplies must be stationed beforehand at appropriate sites around the city, while heavy lifting equipment must be on hand to create routes through rubble-filled streets. As much as possible, the population should be trained beforehand to look after themselves in the hours and days immediately after a quake, leaving the emergency services unhindered in their quest to save lives.

Even in countries that now have earthquake building codes, education is vital to reduce corrupt practices and ensure that the codes are stringently enforced. Failure to tackle these aspects can lead to terrible and unnecessary destruction and loss of life, as demonstrated in the 1999 Izmit quake. Here the collapse rate in quite modern apartment buildings was an almost unprecedented 75 percent in places, largely the result of poor building standards. Due to a huge housing boom in the last few decades, local government capabilities to enforce building codes and control development were overwhelmed, leading to relaxed inspection procedures and dangerously high building densities. The end result was one of the most lethal urban catastrophes in recent years.

Where next?

>>> Californians have been awaiting the "big one" for perhaps 50 years now, and they are still waiting.

> The state's unusually high level of seismic activity—there were almost 12,000 quakes in 1998—is due to movements of the San Andreas Fault and its many subsidiaries. Yet, in the last 100 years, these have resulted in just a single earthquake in excess of magnitude 8. There is no doubt that another quake of this size is due—perhaps within the next few decades. Although it is not exactly clear where it will strike, it will be somewhere on the San Andreas or one of its neighboring faults, which together threaten over 15 million people.

Despite the magnitude 8.25 quake that struck in 1906, San Francisco remains under threat, as does Los Angeles. Here, there have been eight major earthquakes since 565 A.D., spaced at intervals ranging from 55 to 275 years, giving an average return period of 160 years. As it is now 144 years since the last major quake, in 1857, the next could arrive at any moment. Scientists are currently monitoring the swelling and subsidence of the crust at Palmdale, just 30 miles (50km) north of the city, which might reflect severe strain accumulation in advance of a major quake.

Los Angeles sits in what is known as a seismic gap. This is another way of saying that a large quake can be expected at any time, based upon previous return periods. The Great Alaska Earthquake of 1964 filled such a gap. Others exist today, particularly in the Tokyo region and in parts of the Caribbean. Another one can be found just to the south of Istanbul, where a segment of the North Anatolian Fault is widely expected to rupture within the next decade or two. In light of the utter devastation wrought by the 1999 Izmit quake just up the road, the future

for the residents of Istanbul looks far from rosy.

Most people would not even blink at news of a big quake in California, Japan, or Turkey, all of which experience earthquakes on a regular basis. Barring local people, however, few would expect to hear of a catastrophic quake hitting the Seattle region of the U.S. Pacific Northwest. Nevertheless, the seismic threat here is severe, and, even as I write this, a magnitude 6.8 has just struck 40 miles (60km) south of Seattle, causing

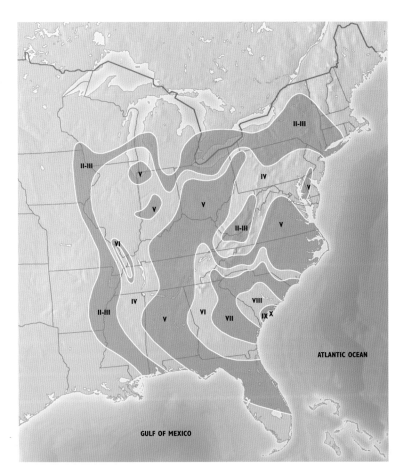

ATLANTIC OCEAN

GULF OF MEXICO

ABOVE Mercalli Scale intensities show that the 1886 Charleston quake in South Carolina was felt nearly 1,200 miles (2,000km) away.

significant damage, although no loss of life. The real threat to the region comes, however, from the gigantic but infrequent quakes that occur offshore in the Cascadia Subduction Zone, where the small Juan de Fuca platelet is plunging down beneath the North American continent. Just 301 years ago, a quake in excess of magnitude 8 generated tsunami that within 15 minutes had inundated the entire coastline with waves up to 50 feet (15m) high. Ten hours later the waves crashed into Japan causing sufficient damage for the event to be recorded for posterity. The next big Cascadia quake is awaited with some trepidation, not only in Seattle and neighboring Tacoma, but also across the border in Vancouver, British Columbia.

To the heart of the plate

Somewhat alarmingly, big earthquakes are not only confined to plate margins. Over long periods of time the strains accumulating at the edges of the plates can insinuate themselves into the hearts of the plates, where they can trigger movements on any available fault. This explains why destructive quakes struck New York in 1884 and Charleston, South Carolina, in 1886. Even in northern Europe, magnitude 7 earthquakes occur, very infrequently, when the crust can no longer contain the strain build-up. A good bet for the next intraplate "big one" is in the New Madrid region, close to Memphis, Tennessee. This area was battered in 1811 and 1812 by three quakes close to magnitude 8, which warped and tilted the crust so severely that the Mississippi river flowed backward for a time. The next New Madrid quake could arrive at any time, causing severe damage not only in Memphis but also in the city of St. Louis in neighboring Missouri.

RIGHT Los Angeles has been damaged before by earthquakes and another big one is on the way.

TARGET JAPAN

Japan is a country that fights a constant battle against natural catastrophes.

Dug into the front line on the western margin of the Pacific's Ring of Fire, the country is regularly battered by typhoons and tsunami and rocked by eruptions from one of the country's 94 active volcanoes. Even worse, Japan is one of the most earthquake-prone regions on the planet, and the 1,800 mile (3,000km) long chain of islands is almost constantly shaken by quakes that release an extraordinary ten percent of the Earth's seismic energy every year.

There is no mystery regarding Japan's perilous geophysical circumstance. Not far from the Pacific shores of the island chain are a series of precipitious deep-sea trenches into which the dense oceanic lithosphere of the Pacific plunges beneath the huge Eurasian Plate that stretches as far west as the center of the Atlantic Ocean. Every time the floor of the Pacific thrusts itself downward into the subduction zone, the crust above is severely shaken, buildings tumble, and people die.

Over the past three centuries Japan has been struck by a serious destructive earthquake about every 30 years. Despite enormous advances in the science of seismology and improvements in building methods, however, there appears to be little evidence to support a reduction in the destructive capacity of future quakes. The big problem is that Japan is an immensely crowded land—its population density is 13 times that of the United States—and because much of the terrain is mountainous and largely uninhabitable, this population is largely concentrated into towns and cities, including a number of huge conurbations. Consequently, strong earthquakes now have much greater targets to aim at and, as shown by the 1995 Kobe quake, the toll they exact—both in terms of economic cost and loss of life—is terrible.

The scale of death and destruction arising from the Kobe tragedy stunned a world that largely believed that such losses only occurred in poor, developing-world countries. Kobe, however, showed that if a quake is big enough then devastation will be the result wherever it strikes. For Japan the big question now is—what will be the result if and when the capital Tokyo is rocked by a big one?

OPPOSITE In 2015, the population of Tokyo will reach a staggering 28 million. How will they fare when the next devastating quake strikes? BELOW Japan is plagued by earthquakes due to the Pacific Plate to the east plunging beneath the Eurasian Plate in the west.

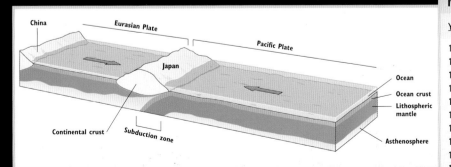

Major Japanese Earthquakes

YEAR	REGION	DEATHS	MAGNITUDE
1662	LAKE BIWA	880	7.2–7.6
1703	OSHIMA	10,367	7.9–8.2
1707	KII PENINSULA	5,038	8.4
1751	TAKADA	1,541	7.0–7.4
1766	AOMORI	>1,200	6.9–7.2
1847	NAGANO	8,174	7.4
1854	NANKAI/TOKAI	6,000	8.4
1923	KANTO (TOKYO–YOKOHAMA)	142,807	8.3
1948	FUKUI	3,769	7.1
1955	KOBE	5,502	6.9

The restless catfish

>>> Although earthquakes still retain the ability to terrify the Japanese population, at least everyone is now familiar with their cause and knows what to do when the violent shaking begins.

> During the time of the Samurai, however, earthquakes were viewed as divine visitations inflicted upon mere mortals by an extraordinary subterranean battle between deity and beast. Earthquakes were thought to be due to a lack of vigilance by Kashima—the protector of the Earth. Kashima's somewhat tiresome occupation was to subdue a lively giant catfish—or *namazu*—using a huge rock. Not surprisingly, Kashima would sometimes tire of this laborious duty and fall into a deep slumber, which gave the catfish a chance to thrash about, causing the Earth above to tremble and shake.

If such a hyperactive catfish did exist, there is little doubt that it would at this moment be flexing its fins beneath the sprawling conurbation of Tokyo and its sister city, Yokohama. There is no question that Japan's capital is a city waiting to die, sited as it is in a region of especially complex and dangerous geology where no fewer than three tectonic plates converge. Here, the great Pacific Plate plunges beneath the Philippine Sea Plate, which in turn dives below the giant Eurasian Plate. The geology is poorly understood, but its hazard implication is more than familiar to the local population. Tokyo–Yokohama is underlain by a number of active faults related to the plate margins that time and time again shake the cities to their foundations. The last time this happened, in the Great Kanto Earthquake of 1923, the result was probably the greatest catastrophe in Japanese history, including the bombing of Nagasaki and Hiroshima. Indeed, images taken of the twin cities after the shaking and the fires had ceased to do their worst were almost indistinguishable from the ruinous legacy of the atomic bombs.

Deceptive calm

In 1923, perhaps 200,000 people died amid the charred and collapsed buildings of the capital and its neighborhood. The intervening years, however, have been remarkable for their seismic calm. Although the Tokyo metropolitan area has been gently to moderately shaken on occasion—most recently by a magnitude 5.9 quake in 1992—it has sustained little damage or loss of life, and there have been no major earthquakes at all. Clearly this cannot and will not last, and the people and the authorities know that some day soon their capital will be

ABOVE In Japanese tradition, a giant catfish —or *namazu*—was held responsible for the earthquakes that regularly struck the country. RIGHT Large, shallow earthquakes have hit all the way down the east coast of Japan since 1920. When and where will the next "Big One" strike?

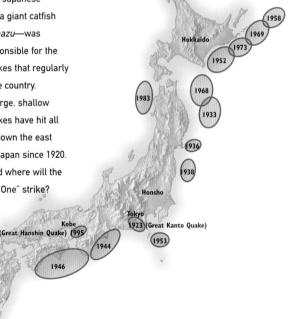

rocked by another giant quake that will bring it to its knees.

The source of the next quake, however, remains uncertain. In fact, Tokyo–Yokohama is threatened by earthquakes from no fewer than four different active faults, movements of which are widely regarded as being overdue. Moderate quakes to the southwest—beneath Suruga Bay and close to the city of Odawara—are thought to be imminent and capable of causing moderate damage to the Tokyo metropolitan area. The biggest threat comes, however, from a repeat of the Great Kanto Quake, out in Sagami Bay to the south, or from a so-called *chokka-gata* quake directly under the city. The Great Kanto

sequel could still be a few decades off but the next *chokka-gata* could arrive tomorrow.

A *chokka-gata* could be as powerful as the quake that wrought terrible destruction upon Kobe in 1995 and there is no reason to suppose that the capital would fare any better. What surprised observers of Kobe was the poor response of the authorities to the aftermath of the quake. For days, blocked roads and railway lines and damaged lines of communications hindered the movement of emergency supplies and equipment, and hundreds of thousands of homeless were left to fend for themselves for as long as a week after the ground stopped shaking. Will Tokyo fare any better?

ABOVE A firefighter lays flowers at a memorial to a colleague killed at Kobe during Japan's last seismic catastrophe.

Meltdown

>>> In California and other quake-prone parts of the planet, early optimism about our ability to predict earthquakes accurately has now vanished, and little research funding is available in this area.

ABOVE Will vigilant Japanese scientists have a warning of the next big one or will they be caught napping?

> Only in Japan is prediction thought to be possible and government officials and city authorities are confident that they will get geophysical warnings from their monitoring systems days—or at least hours—before the next big one. Many scientists believe, however, that there will be no warning. Within seconds the hustle and bustle of a normal day in downtown Tokyo will be transformed into a cacophony of crashing buildings, shattering glass, exploding gas mains, and dying citizens. No precursors, no prediction from the government's Earthquake Assessment Committee, no time to take cover—just violent shaking as if by a gigantic hand, followed by crushing oblivion.

The next Great Kanto Earthquake is likely to register 7 on the Japan Meteorological Agency Scale (JMS) of earthquake measurement, which is as high as it goes. Such an event is described on the scale as "severe damage; over 30 percent of buildings collapse; ground fracturing and landslides common."

Total devastation

This does not, however, even begin to paint a picture of the awful devastation that will result from the coming Tokyo megaquake. Beneath Sagami Bay, 50 miles (80km) to the south of Tokyo, the quake will be triggered as one great tectonic plate thrusts itself almost instantaneously beneath another, sending violent shock waves hurtling northward toward the capital. Within a few seconds the inhabitants will notice a low, rumbling vibration that, in no time at all, will translate into an earth-shattering roar as the Earth starts to roll like the surface of the sea and buck like a rodeo bronco.

In the older parts of the city, traditional buildings with their heavy tiled roofs and weak walls will swiftly crash to the ground, killing and maiming those inside. Downtown, panic will quickly give way to terror as glass and masonry from hundreds of office buildings crash to the ground and detached giant advertising billboards crush pedestrians and cars below. As drivers leave their vehicles to search desperately for cover, the roads will become gridlocked and the situation will worsen as weaker structures start to collapse and power lines are brought down. Along the waterfront and in much of the downtown area, buildings will topple and sink out of sight as the soft sediment on which they are constructed begins to liquefy and behave as water. The terrible shaking will probably end after a minute or so, but this will not be the end of the destruction and loss of life.

The explosive jolt to the seabed out in Sagami Bay will have shaken up the ocean and organized it into a series of devastating tsunami. Barely minutes after the shaking ceases, wave after wave will pound the factories, port facilities, and oil refineries along the coast, reducing them to rubble and igniting great conflagrations as oil pours from damaged tanks. Particularly if the weather is warm and dry with a strong wind, fire will soon take hold, fed by exploding gas tanks, fractured gas mains, and oil and chemical spills. Over a million wooden buildings will provide the fuel, as a thousand small fires seek one another out to form great conflagrations the size of entire city blocks. Tens of thousands will be trapped and burned alive as they try to escape the fast-moving firestorms.

The great fires may last for days, leaving little untouched when the flames are finally quenched. A death toll in the hundreds of thousands will be no surprise, while current estimates by the insurance market expect economic losses to total $7 trillion. This is 35 times greater than Kobe, the most costly natural disaster to date, and 70 percent of the current gross domestic product of the country. The resulting economic shock waves will hurtle across the planet bringing economy after economy to its knees and, perhaps, triggering global economic meltdown. With over 20 million inhabitants, the Tokyo region is one of the planet's biggest economic engines: many believe that when Tokyo stops working so will the rest of the world.

RIGHT The Great Kanto Quake of 1923 obliterated the Japanese capital and may have killed up to 200,000 people.

Japan Meteorological Agency Scale (JMS)

INTENSITY ON JAPAN METEOROLOGICAL SCALE	EFFECTS AND DAMAGE	APPROXIMATE EQUIVALENT ON MODIFIED MERCALLI SCALE
0	NOT FELT: DETECTABLE ONLY BY SEISMOGRAPHS	I
1	FAINT TREMOR FELT ONLY BY PEOPLE AT REST	II
2	NOT DETECTABLE BY ALL: LAMPS SWING AND WATER IN BOWLS IS DISTURBED: VIBRATION SIMILAR TO A PASSING TRUCK	III
3	DETECTABLE BY ALL: HOUSES SHAKE, DOORS CREAK: WINDOWS RATTLE, CROCKERY DISTURBED	IV–V
4	SLIGHT DAMAGE: FELT BY ALL: HOUSES SHAKE VIOLENTLY: LOOSE PLASTER MAY FALL AND HEAVY FURNITURE MOVED	VI
5	NOTABLE DAMAGE: NOTICED BY PEOPLE DRIVING CARS: WALLS CRACK, CHIMNEYS DAMAGED	VII
6	MAJOR DAMAGE: FEWER THAN 30% OF BUILDINGS COLLAPSE: GROUND FRACTURING AND LANDSLIDES OCCUR: MOST PEOPLE NOT ABLE TO STAND	VIII–IX
7	SEVERE DAMAGE: OVER 30% OF BUILDINGS COLLAPSE: GROUND FRACTURING AND LANDSLIDES COMMON	X–XII

Living on borrowed time

>>>There is an important distinction between scaremongering and enlightenment and I have tried in this book to educate about the tectonic threat to the human race rather than to terrorize the reader.

>The first step toward mitigating the impact of a natural hazard lies in raising awareness of the risk and improving understanding of the phenomena. If I have managed to go some way toward this then I am reasonably content. Specialized texts provide a wealth of more detailed information on earthquakes, tsunami, and volcanic eruptions, but I hope in this overview to have encapsulated the main features and addressed the principal issues relating to the impact of these natural forces on human activities.

Unquestionably, the new century will bring new challenges. As the global population continues to rise and become increasingly urbanized, so the impact of tectonic hazards will escalate inexorably. This is now recognized and expertise in risk science and disaster management is being mobilized to try and counter the growing threat. Success, however, will be enormously difficult to achieve, because

the forces that are increasing vulnerability, particularly in developing countries, are overwhelming. The battle for prosperity and living space conspire together to construct an almost insurmountable barrier to natural hazard reduction, by engendering poor quality building in high risk areas. Not until governments and authorities can be encouraged to give public safety a higher priority will even a small dent be made in the growing casualty figures.

The threat to our future

What then will be the scale of the tectonic threat that our children and grandchildren will face? Over the next 100 years, earthquakes will shake our planet some 50 million times, with over 10,000 of these having the potential to cause serious damage and loss of life. Around 2,000 magnitude 7 and a few hundred magnitude 8 quakes will pepper the major seismic zones coincident with the margins of

OPPOSITE Within the next 100 years, Tokyo (shown) and Los Angeles are both likely to be struck by major destructive earthquakes.
OPPOSITE BELOW So far, we have survived the bomb, but how will we fare against the worst Mother Nature can throw at us?

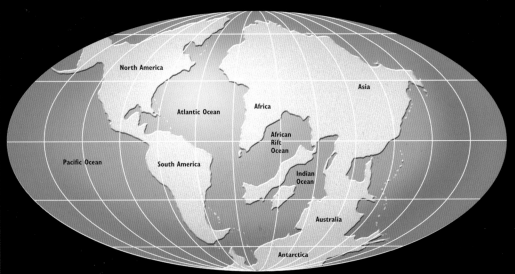

LEFT In 150 million years' time, movements of the Earth's tectonic plates will bring earthquakes and volcanoes to new parts of the planet, such as the eastern U.S. and Australia.

North America

Asia

Africa

Atlantic Ocean

African
Rift
Ocean

Pacific Ocean

South America

Indian
Ocean

Australia

Antarctica

the tectonic plates. Both Tokyo and Los Angeles are likely to be struck by severe earthquakes, while the final strand of the North Anatolian Fault will rupture, bringing Istanbul to its knees. Perhaps for the first time 1 million people will die as a powerful quake obliterates a seething metropolis in India, Mexico, or China. Maybe another great Cascadia earthquake will tear the crust off the coast of the U.S. Pacific Northwest, sending huge tsunami crashing into Seattle.

Volcanoes will continue to explode into life and we can expect as many as 5,000 eruptions before the century is out. Undoubtedly the old favorites, Etna and Stromboli in Italy and Kilauea in Hawaii, will continue to rumble on, but predicting where the next really big bangs will occur is very much like playing the lottery. With an ominous silence reigning since Vesuvius last erupted 55 years ago, there is a good chance that the volcano will once again vent its spleen on the city of Naples, where some 800,000 people are under threat. Elsewhere, keep an eye out for new eruptions at Mount Rainier in Washington State, at Mammoth Mountain in California, and on the Caribbean island of Dominica.

In 2100, will our race greet the new century from the perspective of a technologically advanced, affluent global society or as the bruised and battered remnants of a race devastated by a volcanic super-eruption or a giant tsunami? Even if it is the former, doom and disaster will continue to dog the years ahead. The tectonic plates will continue to move, changing over millions of years the familiar shapes of the continents. Earthquakes will devastate parts of the Earth that have not experienced them before, while new volcanoes will spring up in regions once regarded as safe from Vulcan's wrath. And, in the long term, it is virtually certain that global tectonic catastrophe will strike within the next 50,000 years.

Glossary

Accretion the process of accumulation of space debris that formed the Earth and other planets

Aerosol microscopic droplets

Albedo the reflectiveness of the Earth's surface

Asthenosphere the partially molten layer of the upper-most mantle

Basalt a dark, high-density volcanic rock that floors the ocean basins

Caldera a giant volcanic crater formed by collapse of the Earth's crust

Conduction the mechanism of heat transfer in solids

Conservative plate margin a plate boundary where two plates slide past one another

Constructive plate margin a plate boundary where new lithosphere is formed

Convection the mechanism of heat transfer in fluids

Destructive plate margin a plate boundary where old lithosphere is consumed

Earthquake storm a sequence of earthquakes clustered in space and time

Effusive the term used to describe lava-dominated eruptions

Epicenter the point on the surface directly above an earthquake source

Flood basalts highly voluminous eruptions of low-viscosity basalt magma

Focus the source of an earthquake

Granite a pale, low-density magmatic rock that forms the bulk of the continents

Lahar a torrent of volcanic debris

Liquefaction the phenomenon whereby unconsolidated material acts as a fluid

Lithosphere the rigid outer layer of the Earth

Mantle the region between the crust and the core of the Earth

Microseisms tiny earthquakes

Mid-Ocean Ridge the topographic rise in an ocean basin where new magma reaches the surface

Modified Mercalli Scale a scale that classifies a quake according to the damage it causes rather than its strength

Nuée ardente see Pyroclastic flow

Olivine a green mineral common in the mantle

P wave a seismic wave that repeatedly squeezes and stretches the rock it passes through

Pyroclastic flow hurricane blasts of superheated gas, pumice, and ash

Retrofitting upgrading old buildings to make them quake-proof

Return period the characteristic frequency of a tectonic hazard

Richter Scale a scale that measures the size of earthquakes

Seismic electrical signals electrical signals in the crust that are claimed to precede earthquakes

Seismograph a device for recording earthquakes

Silica a molecule made up of silicon and oxygen and the most common constituent of most rocks

Strain meter a device that measures strain accumulation in rocks

Stress transfer the mechanism by which movement on one fault increases the stresses acting on another

Subduction zone a deep ocean trench within which old lithosphere is consumed and melted

Surface wave a seismic wave that causes the surface to roll like the surface of the ocean

S wave a seismic wave that causes violent sideways shaking

Tectonic plate giant, mobile, rocky slabs that make up the Earth's outermost rigid layer

Tephra the collective term for volcanic debris ejected into the atmosphere

Tsunami giant sea waves normally generated by earthquakes but also by collapsing volcanoes

Volcanic Explosivity Index a scale that measures the violence of a volcanic eruption

Volcanic winter the dramatic fall in temperature that follows a volcanic super-eruption

Index

Credits

Quarto would like to thank and acknowledge the following for supplying pictures reproduced in this book:

Key: l left; r right; c center; t top; b bottom

The Art Archive pp 34 (Palazzo Barberini/A Dagli Orti), 44-45 t (Hermitage Museum, Saint Petersburg/A Dagli Orti), 69 (Album/ Joseph Martin), 104 (Dagli Orti); **Art Directors** pp 6-7 (Viesti Collection), 10-11c (Th-Foto Werbung), 36 & 42 (Viesti Collection), 61 (M Barlow), 92 (Viesti Collection), 112 (Dinodia), 122 (Viesti Collection), 123 (Trip); **BBC** 88-89c (Tristan Marshall); **British Geological Survey © NERC** p 111; **©Corbis** pp 13b David Turnley, 17 (Raymond Gehman), 19 (Ralph White), 33t (Gianni Dagli Orti), 49 (digital image ©1996 Corbis; original image courtesy of NASA/Corbis, 72 (Brian Enting), 94 b, 95 & 98 (Bettmann) 99 l & r (Hulton-Deutsch Collection, 121 l (George Hall), 121 r (Bettmann), 128 (Annie Griffiths Belt), 135 (Michael S Yamashita), 137 (Hulton-Deutsch Collection); **Sylvia Cordaiy Photo Library** pp 4-5(Guy Marks), 40-41 (Guy Marks); **Simon Day** p89; **John Grattan** p70; **Bill McGuire** pp 33b, 35, 51, 52r, 54l, 55r, 56 br, 67 t, 75, 75, 76, 77t, 86-87c, 87r, 115t; **NASA** pp 32b (GSFC/MITI/ERSDAC/JAROS); **NGDC, Boulder, C0,** pp 15 (University of Colorado), 27 (Clifford E Harwood, Encino, CA), 58 (R L Rieger, US Navy), 59 (R P Hoblitt, US Geological Survey), 62 (NOAA/NGDC), 63 (University of Colorado), 96 b (NOAA/NGDC), 97 t (G K Gilbert, US Geological Survey), 97 b (University of California, Berkeley), 107 & 113 (NOAA/NGDC), 124 l (M R Mudge, US Geological Survey), 125 l (R E Wallace, US Geological Survey), 126 (University of Colorado), 127b (National Geophysical Data Center), 133 (University of Colorado), 137 t (H C Shah, Stanford University); **NOAA** p119; **National Geographic Society**, ©Institute for Exploration p29; **Panos Pictures** pp23, 102-3 bc (Jeremy Horner), 109 b (Jim Holmes); **Pictor International** pp 22, 84, 90-91, 115 b, 131, 132; **Pictures Colour Library** p 139t; **Frank Spooner Pictures** pp 9 (Laski), 44-45b (M Deville), 50, 57 (H.Tazieff), 58 (R L Rieger), 64 (Gilles Bassignac), 65b, 73t (Andrew Reid), 93 (Erick Bonnier), 100b (Hosaka), 101 t (Naoto Hosaka), 101b (Naoto Hosaka), 102 bl (ABC Basin Ajansi), 105 b (Roger-Viollet), 106-7 tc (Giboux), 116 (Remi Benali), 117 (Remi Benali), 120 (Kaku Kurita), 127t (Giboux), 136 (Marc Gantier), 139b (Edwards); **Steven Ward** p88; **The Syndics of Cambridge University Library** pp 54-5 c (Royal Commonwealth Society Collections); **Dr Eysteinn Tryggvason** pp 14, 37, 81; **University of Iceland, Department of Geology, Thorarinsson collection**, pp 46, 47; **US Geological Survey** pp 2, 11r(HVO) , 52 l (HVO), 53 (HVO), 106 tl (Earthquake Information Bulletin 297); **Wilhamina F Jashemski** Courtesy Aristide D Caratzas, Publisher. p45br.

All other photographs and illustrations are the copyright of Quarto Publishing plc. While every effort has been made to credit contributors, we would like to apologize in advance if there have been any omissions or errors.